Key Challenges in Geography

EUROGEO Book Series

This book series addresses relevant topics in the wide field of geography, which connects the physical, human and technological sciences to enhance teaching, research, and decision making. Geography provides answers to how aspects of these sciences are interconnected and are forming spatial patterns and processes that have impact on global, regional and local issues and thus affect present and future generations. Moreover, by dealing with places, people and cultures, Geography explores international issues ranging from physical, urban and rural environments and their evolution, to climate, pollution, development and political economy.

Key Challenges in Geography is an initiative of the European Association of Geographers (EUROGEO), an organization dealing with examining geographical issues from a European perspective, representing European Geographers working in different professional activities and at all levels of education. EUROGEO's goal and the core part of its statutory activities is to make European Geography a worldwide reference and standard. The book series serves as a platform for members of EUROGEO as well as affiliated National Geographical Associations in Europe, but is equally open to contributions from non-members.

The book series addresses topics of contemporary relevance in the wide field of geography. It has a global scope and includes contributions from a wide range of theoretical and applied geographical disciplines.

Key Challenges in Geography aims to:

- present collections of chapters on topics that reflect the significance of Geography as a discipline;
- provide disciplinary and interdisciplinary titles related to geographical, environmental, cultural, economic, political, urban and technological research with a European dimension, but not exclusive;
- deliver thought-provoking contributions related to cross-disciplinary approaches and interconnected works that explore the complex interactions among geography, technology, politics, environment and human conditions;
- publish volumes tackling urgent topics to geographers and policy makers alike;
- publish comprehensive monographs, edited volumes and textbooks refereed by European and worldwide experts specialized in the subjects and themes of the books;
- provide a forum for geographers worldwide to communicate on all aspects of research and applications of geography, with a European dimension, but not exclusive.

All books/chapters will undergo a blind review process with a minimum of two reviewers.

An author/editor questionnaire, instructions for authors and a book proposal form can be obtained by contacting the Publisher.

The Key Challenges in Geography Series is now indexed in SCOPUS.

Rafael de Miguel González · Pilar Rivero Gracia
Editors

Geospatial Technologies for Heritage Education

Springer

Editors
Rafael de Miguel González
University of Zaragoza
Zaragoza, Spain

Pilar Rivero Gracia
University of Zaragoza
Zaragoza, Spain

ISSN 2522-8420 ISSN 2522-8439 (electronic)
Key Challenges in Geography
ISBN 978-3-031-96371-1 ISBN 978-3-031-96372-8 (eBook)
https://doi.org/10.1007/978-3-031-96372-8

This work was supported by EUROGEO ivzw.

This Springer imprint is published by the registered company Springer Nature Switzerland AG
The registered company address is: Gewerbestrasse 11, 6330 Cham, Switzerland

Contents

Geospatial Technologies for Heritage Education

Rafael de Miguel González and Pilar Rivero Gracia

Abstract This book addresses the advantages of using geospatial information technologies in education for natural and cultural heritage, and sustainable territorial development, as well as the challenges posed by the growing use of artificial intelligence in the face of the necessary deepening of critical, autonomous, and personal thinking by the student. In this way, in addition to this introduction the different authors have chosen fourteen different approaches that aim to enrich the conceptual debate, but also the methodological diversity in the teaching of heritage for its knowledge, conservation, and promotion by future generations.

Geospatial Competences in School Education

Incorporating geospatial competences into school curricula equips students with essential skills for understanding and engaging with the world around them. Geographic Information Systems (GIS), remote sensing, and other geospatial tools enable students to explore spatial relationships, analyze data, and develop critical thinking abilities. By learning to interpret spatial information and visualize geographic data, students become more informed citizens capable of addressing complex global challenges such as climate change, urban planning, and cultural preservation.

R. de Miguel González (✉) · P. Rivero Gracia
ARGOS Research Group, IUCA-Unizar, Faculty of Education, University of Zaragoza, Zaragoza, Spain
e-mail: rafaelmg@unizar.es

P. Rivero Gracia
e-mail: privero@unizar.es

R. de Miguel González and P. Rivero Gracia (eds.), *Geospatial Technologies for Heritage Education*, Key Challenges in Geography,
https://doi.org/10.1007/978-3-031-96372-8_1

Developing Spatial Literacy

Spatial literacy, or the ability to interpret and analyze spatial data, is foundational for understanding geography and heritage. Students learn to navigate maps, overlay datasets, and examine spatial distributions of cultural heritage sites. This literacy is not limited to reading traditional maps but extends to understanding digital spatial data and interactive visualizations.

For instance, students can use GIS to analyze archaeological findings, visualize ancient trade routes, or assess urban expansion's impact on heritage preservation. Learning how to interpret satellite imagery and topographic maps helps students grasp concepts like land use, environmental change, and resource distribution. By engaging in exercises that require the analysis of spatial patterns, students refine their problem-solving skills and enhance their ability to draw conclusions from complex data sets.

Digital Heritage Education

Digital tools like GIS facilitate interactive learning experiences, allowing students to "travel" through time and space. Virtual reconstructions of historical landscapes, created using GIS data and 3D modeling, make it possible for students to visualize how environments and communities evolved over centuries. These digital environments not only make history more engaging but also deepen students' understanding of historical processes and cultural transformations.

Such tools also promote spatial citizenship by encouraging students to consider their roles in preserving cultural and natural heritage. Through digital simulations, students can explore the effects of human decisions on the environment and heritage preservation. This approach fosters a sense of responsibility and encourages proactive participation in cultural conservation. For example, students might simulate the impact of urban development on local heritage sites, prompting discussions about sustainable growth and cultural stewardship.

Citizen Science and Collaborative Cataloging

Geospatial technologies empower students to participate in citizen science projects, contributing to the collaborative cataloging of cultural heritage. By using GIS applications to map local landmarks, record historical narratives, or document endangered sites, students gain hands-on experience while contributing to communal heritage preservation efforts.

In addition, students can engage in projects that involve surveying their local communities, collecting data on historic sites, and contributing to open-source mapping platforms. These activities not only enhance students' technical skills but also foster community engagement and pride. By documenting stories and landmarks, students help create a shared historical record that benefits researchers, educators, and the broader public.

Space, Time, Landscape, and Heritage

Understanding heritage requires examining the interconnected dimensions of space, time, and landscape. Geospatial technologies provide the tools to analyze these dimensions in a manner that is engaging, comprehensive, and educational. Students learn how landscapes shape and are shaped by cultural practices, environmental changes, and historical events. This second part of the book faces this topic throughout four chapters.

Landscape Reconstruction and Spatial Analysis

GIS and remote sensing technologies enable the reconstruction of historical landscapes, revealing the spatial organization of ancient societies. For example, LiDAR scanning has uncovered lost settlements and infrastructure in dense environments, offering insights into the interactions between human activity and natural landscapes. These reconstructions not only aid research but also serve as valuable educational resources.

By examining reconstructed landscapes, students gain insights into how ancient civilizations adapted to their environments. They learn about agricultural practices, settlement patterns, and trade networks. This understanding fosters a deeper appreciation of the resilience and ingenuity of past societies. Additionally, students can use GIS to conduct their own spatial analyses, such as studying the proximity of ancient settlements to natural resources or transportation routes.

Evaluating Heritage Through Space and Time

Cultural and natural heritage must be understood within the contexts of their temporal and spatial dynamics. GIS tools allow students to analyze how landscapes have been shaped by human and natural forces over centuries. For instance, mapping the spread of the migration of cultural artifacts across continents can provide a richer understanding of global history.

By examining historical maps and satellite imagery, students can trace environmental and cultural changes over time. They can investigate how deforestation, urbanization, and climate change have impacted heritage sites.This analysis not only enhances historical understanding but also raises awareness about contemporary conservation challenges.

Cross-Border Cultural Cooperation

The spatial dimension of heritage often transcends national boundaries, necessitating cross-border cooperation. GIS facilitates this collaboration by creating shared platforms for data collection and analysis, enabling students and educators to work on joint heritage projects across regions.

Through international partnerships, students can collaborate with peers in other countries to map transboundary heritage sites, compare historical narratives, and share research findings. This global perspective fosters cultural understanding and empathy, highlighting the interconnectedness of human history. It also provides opportunities for language learning, interdisciplinary research, and digital literacy development.

Uncovering Hidden Heritage

Geospatial tools are instrumental in identifying and preserving "hidden heritage"—sites or artifacts that are obscured by environmental or urban development. Through remote sensing and spatial analysis, students can explore the challenges of conserving heritage in danger, fostering a deeper appreciation for cultural preservation.

For instance, students can study the use of ground-penetrating radar and drone technology to discover buried structures. They can analyze case studies where these technologies have been used to protect endangered heritage, learning about the importance of documentation and advocacy. This approach encourages students to think critically about the value of heritage and the ethical considerations involved in preservation efforts.

Inquiry-Based Problem Learning for Teaching Geography and History

Inquiry-based problem learning (IBPL) places students at the center of the educational process, encouraging them to investigate and solve real-world problems. Geospatial technologies are particularly suited to this pedagogical approach, offering dynamic tools to explore geography and history.

Promoting Critical Thinking Through Problem Solving

Using GIS, students can address questions such as, "How did natural features influence the location of ancient cities?" or "What are the effects of modern urbanization on historical sites?" These exercises require students to gather data, formulate hypotheses, and test their ideas, fostering critical and analytical thinking.

By engaging in problem-solving activities, students develop skills in data interpretation, hypothesis formulation, and evidence-based reasoning. These skills are transferable to a wide range of academic and real-world contexts, preparing students for future academic pursuits and careers.

Simulating Historical and Geographical Scenarios

Game theory and simulation tools integrated with GIS offer unique opportunities for experiential learning. For example, students can role-play as planners negotiating the conservation of a UNESCO World Heritage Site, balancing the interests of stakeholders such as local governments, businesses, and preservationists. This approach encourages collaboration and ethical reasoning.

Students learn to consider multiple perspectives, weigh competing priorities, and make informed decisions.They also develop communication and negotiation skills, which are essential for collaborative problem-solving in professional contexts.

Exploring Historical Narratives

By overlaying historical maps on contemporary data, students can explore historical narratives with spatial precision. GIS allows for the visualization of events such as migration patterns, battles, or trade networks, bringing historical lessons to life. For example, mapping the spread of the Roman Empire using GIS can help students understand the interplay of geography, politics, and culture.

This method encourages students to think critically about historical causality and complexity. They learn to ask questions about why certain events occurred, how they unfolded, and what consequences they had for different societies. This deepens their historical understanding and fosters a more nuanced appreciation of the past.

Addressing Contemporary Challenges

IBPL with geospatial technologies also prepares students to tackle contemporary issues, such as climate change's impact on heritage sites or urban development's encroachment on archaeological areas. This approach aligns heritage education with broader societal challenges, making learning relevant and impactful.

Students develop skills in sustainability analysis, policy evaluation, and community engagement. They learn to propose solutions to complex problems and to consider the long-term implications of their decisions. This empowers them to become active, responsible citizens who contribute positively to their communities.

Social Studies Teacher Training from a Multidisciplinary Approach

To effectively integrate geospatial technologies into heritage education, teachers must receive comprehensive training that combines technical skills with interdisciplinary insights. Such training ensures educators are well-equipped to inspire and guide their students.

Geospatial Tools in Teacher Training

Teacher training programs must emphasize the practical application of GIS, remote sensing, and other tools. Workshops and hands-on exercises can familiarize teachers with the software and methodologies needed to create engaging lessons. For instance, teachers can learn to design GIS-based activities that analyze local heritage sites or model historical changes in their communities.

Multidisciplinary Collaboration

Heritage education benefits from a multidisciplinary approach that integrates history, geography, archaeology, and environmental science. Teacher training should reflect this integration, providing educators with the knowledge to draw connections between disciplines. For example, teachers might learn how to use GIS to study the ecological impact of historical agricultural practices.

Addressing Heritage in Danger

Training programs should also prepare teachers to address issues of heritage in danger, such as sites threatened by conflict, neglect, or natural disasters. Geospatial technologies play a critical role in these efforts, enabling the documentation and monitoring of vulnerable sites. Teachers can incorporate these themes into their curricula, raising awareness among students about the importance of heritage preservation.

Fostering Spatial and Analogical Thinking

By promoting spatial and analogical thinking, teacher training programs ensure educators can effectively convey the relationships between geography, history, and heritage. Spatial thinking enables students to understand the "where" and "why" of historical events, while analogical thinking helps them draw connections between past and present.

Conclusion

Geospatial technologies are reshaping heritage education by providing dynamic tools for exploration, analysis, and collaboration. From equipping students with geospatial competencies to fostering inquiry-based problem solving, these technologies empower educators to create engaging, interdisciplinary learning experiences. By training teachers in the practical application of GIS and promoting a holistic understanding of space, time, and heritage, geospatial technologies ensure that heritage education remains relevant in a rapidly changing world. Through these efforts, students and educators alike become stewards of cultural and natural heritage, preserving it for future generations while deepening their understanding of the past.

This book is the main result of the research project *Digital Competences, Learning Processes, and Cultural Heritage Awareness: Quality Education for Sustainable Cities and Communities*. In addition to this introduction, it comprises fourteen chapters that delve into the topics previously described, authored primarily by researchers who are members of the ARGOS reference research group, or to the research network *Red-14*, both in social sciences education (*didáctica de las ciencias sociales*). However, the inclusion of this book in EUROGEO's current editorial collection, *Key Challenges in Geography*, has allowed for the incorporation of other European research studies. These studies, conducted outside of Spain, focus on the same lines of integrating geospatial technology for heritage education.

Last, in a determined effort to maximize scientific impact, this book has been published in open access to facilitate the dissemination of knowledge and the results of the research conducted on this topic. For this reason, the editors of this book, together with the authors, wish to express their gratitude to the main source of funding that made this volume possible: national research project I + D + I PID2020-115288RB-I00. *Competencias digitales, procesos de aprendizaje y toma de conciencia sobre el patrimonio cultural: educación de calidad para ciudades y comunidades sostenibles*, funded by the Spanish Ministry of Science, MICIU/AEI/

https://doi.org/10.13039/501100011033/. Beside this, it is also part of the lines of the Reference Research Group *ARGOS, Investigación en Didáctica de las Ciencias Sociales*, S50_23R, funded by the Government of Aragon, and of the network RED 2022-134252-T *Red 14 de investigación en didáctica de las ciencias sociales*, funded by the Spanish Ministry of Science, too, MICIU/AEI/https://doi.org/10.13039/501 100011033/.

Geospatial Analysis of Prehistoric Cultural Heritage Sites: Salbyk Valley in Khakassia Case Study

R. Paranin, A. Paranina, L. Marsadolov, and T. Andreeva

Abstract This chapter shows the possibilities of geospatial analysis of cultural heritage sites based on the use of remote research methods and new software for information processing. The features of the current stage of research of the sacred complex of the Salbyk Valley in the Republic of Khakassia are considered. The features of the geolocation and structure of objects that ensure the implementation of instrumental functions in the system of ancient navigation are analyzed. The proposed research methods can be implemented during student field practices, and can be used to obtain new knowledge about heritage sites in other regions.

Keywords Republic of Khakassia · Salbyk archaeological complex · Cultural heritage · Sacred object · Navigation · Remote sensing methods · GIS

Introduction

Prehistoric monuments of the cultural heritage of the Salbyk Valley in Khakassia are still included in the living folk tradition, and also attract the attention of researchers and tourists. In the first half of the 1st millennium BC, more than 100 burial mounds were erected in Salbyk, attributed to the Tagar archaeological culture of the "Scythian time", of which 15 burial mounds are large. Near the burial mounds are more ancient objects—steles and menhirs of the Bronze Age. Research of this complex over several decades was carried out by the Sayano-Altai Archaeological Expedition of the State Hermitage Museum (SAAE SH). Astroarchaeological and geoarchaeological aspects

R. Paranin · A. Paranina (✉) · T. Andreeva
Herzen State Pedagogical University of Russia, St. Petersburg, Russia
e-mail: galina_paranina@mail.ru

L. Marsadolov
The State Hermitage, Saint-Petersburg, Russia
e-mail: marsadolov@hermitage.ru

T. Andreeva
Saint-Petersburg State University, St. Petersburg, Russia

R. de Miguel González and P. Rivero Gracia (eds.), *Geospatial Technologies for Heritage Education*, Key Challenges in Geography,
https://doi.org/10.1007/978-3-031-96372-8_2

were considered with the participation of employees of the Central Astronomical Observatory of the Russian Academy of Sciences (CAO RAS) and the Faculty of Geography of The Herzen State Pedagogical University of Russia (Marsadolov 2010; Marsadolov et al. 2023; etc.). Using the author's materials from field and office studies of these objects as an example, one can consider the effectiveness of modern methods of geospatial analysis.

A feature of geographical research methods is their focus on identifying spatial connections in territorial systems at local, regional, and global levels. For ancient cultural heritage sites, geospatial analysis allows one to determine information functions (including navigational ones). The use of remote sensing materials and modern software for image processing opens new opportunities for extracting useful information, refining the interpretation of objects, and presenting them to the general public: improving existing scientific paleoreconstructions, improving the quality of information support for humanitarian education and educational tourism.

If in the mid-1990s, no more than 1000 people visited the Salbyk Valley per year, in the early 2000s—up to 10,000, then after 2009, more than 30,000 visitors have been recorded. In 2007, a municipal open-air museum, "Ancient burial mounds of the Salbyk steppe", was created here.

Objects of the Salbyk Valley and Geographical Research Methods

The Salbyk Valley is located at the foot of the Kuznetsk Alatau Mountains, 60 km northwest of the city of Abakan (the capital of Khakassia). Of the many cultural heritage sites in the valley, the most famous are the Bolshoi Salbyk Kurgan, the surrounding burial mound complex, and the Salbyk Gate. The Great (Bolshoi) Salbyk Kurgan (burial mound) is one of the largest megalithic monuments of the ancient nomadic tribes of Southern Siberia (the height of the embankment before excavations is 11.5 m). The highest and most revered "fertility" stone is located at the entrance to the mound. The phallic shape, the color of the material, and the high degree of weathering indicate that this stone is older than the mound and was previously part of a Bronze Age sanctuary from the 3rd–2nd millennia BC (Fig. 1a).

The Salbyk Valley burial mound field contains over 100 burial mounds of the Tagar culture of the Scythian period, 15 of which have a diameter of over 50 m. The geometry of the burial mounds is noteworthy—they form linear "chains", often parallel. The most expressed direction is north-west, some "chains" of burial mounds record the directions of the southern line (N–S) and the geographic parallel (W–E). The gates of the burial mounds face east on the western side of the large burial mounds, there are menhirs, sometimes steles with faces. The Salbyk Gates—two menhirs symbolizing a man and a woman (Fig. 1b)—are located 6 km north-east of The Great Salbyk Kurgan (burial mound)—in the saddle of The Two-rock Ringe (one of the spurs of The Batenevsky Ridge). The saddle rises above the dry steppe plains by

Fig. 1 Objects of Salbyk Valley: **a** The Great Salbyk Kurgan (burial mound), fertility stone; **b** Salbyk Gate. Photos by R. V. Paranin

4–6 m and is a corridor accessible for overcoming the ridge using wheeled transport. The site offers wide panoramas with silhouettes of mountains on the horizon. The local population uses menhirs in religious and cult practices, as evidenced by the nearby stone pyramids and ribbons on the bushes of steppe acacia (*Caragana frutex*).

Field research of the Salbyk Valley objects was carried out during long-term comprehensive practice (2021) with the participation of third year students of the

Faculty of Geography (doing undergraduate program in the fields of Geography and Geographical Education). The following types of work were carried out: (1) photo recording of objects; (2) leveling survey of the locations of monuments; (3) remote sensing of monuments and the surrounding landscape. A planned survey of The Great Salbyk Kurgan and the Gates of the Salbyk Valley of the Kings was made from a height of 50 m (Dji Mavic 1 quadcopter), the image was compared with a topographic plan and satellite images (Kiselev 1956; https://www.google.ru/intl/ru/earth/).

Modern software (QGIS, Pixelmator Pro), information resources (Google Earth, Sentinel images, Bing, etc.), and standard geographical methods (landscape and compass survey with leveling elements) were used to process the materials. Geotopological analysis of the space containing cultural heritage objects was carried out using the method of Professor A. N. Lastochkin (https://www.livelib.ru/author/347337/top-a-n-lastochkin?ysclid=m1pktohj1v795422942): the boundaries of geotopes (locations) were structural lines running along the points of maximum curvature of the surface (in plan—keel and ridge lines; in profile—edges and back seams).

Discussion of Results: Geospatial Methods of Analysis—A Source of New Information About Cultural Heritage Monuments

The importance of modern technical means and geographical methods in interpreting the information preserved by archaeological objects is revealed in the context of the long-term history of their research. Three main stages can be distinguished for the objects of the Salbyk Valley: the eighteenth—late twentieth centuries. Classical—research using archaeological methods; 1990–2010. Complex—work involving methods of mathematics and natural sciences (astronomy, geology, geography); since 2010, modern—using the navigation concept of modeling geocultural space, new technical and software tools for spatial analysis. Each of the listed stages was based on the achievements of the previous time and made an important contribution to the evolution of scientific ideas about the creators of this complex—the ancient nomads of Siberia.

Classic Stage in the Exploration of the Great Salbyk Kurgan

The first mentions in scientific literature date back to the mid-eighteenth century. Historian G. F. Miller (1705–1783) first described and measured it in 1739, and then included this mound in the "History of Siberia" (published in 1750, reprinted by Miller 1937, etc.). Local residents told him that the previously conducted (robber) excavations were unsuccessful—12 people did not find any valuables in it in

1.5 months of work (Paranina 2011). The famous Siberian researcher I. T. Savenkov studied this mound in 1899 and left a commemorative inscription on the largest south-eastern stone of the fence. In June 1910, N. V. Fedorov photographed this mound and the largest vertical steles, and measured the stones protruding from the embankment. In 1939, archaeologist S. V. Kiselev (1905–1962) examined the burial mound and the Salbyk Valley during an archaeological reconnaissance expedition of the State Historical Museum and the State Academy of the History of Material Culture. The Great Salbyk Kurgan was studied in 1954–1956 by the Khakassian Archaeological Expedition of the Institute of the History of Material Culture of the USSR Academy of Sciences under the leadership of Moscow archaeologist Kiselev (1951, 1956). S. V. Kiselev's wife, archaeologist L. A. Evtyukhova, and young Moscow archaeologists M. A. Devlet, S. V. Zotova, N. N. Terekhova, E. A. Novgorodova, V. V. Volkov, and L. R. Kyzlasov, who later became well-known specialists, participated in the field expedition work. New methods of that time: study of the mound embankment "for demolition"; cleaning and sketching of all the fence slabs; detailed graphic sketching and photo recording of objects. The few objects found during the excavations of The Great Salbyk Kurgan, after temporary study at the Institute of the History of Material Culture of the USSR Academy of Sciences were transferred for permanent storage to the Abakan Museum in Khakassia and the State Historical Museum in Moscow. After completion of the excavations, The Great Salbyk Kurgan was transferred to the museum, the fence slabs were cleared from the inside, and the eastern "entrance" was cleared of soil.

A detailed analysis of the burial rite, stone structures of Salbyk and the Tagar culture was carried out by Devlet (1976) and other specialists. Excavations of objects of the ancient quarry in Khyzyl-Khaya were carried out by Sunchugashev (1989) is a specialist in mine workings and irrigated agriculture monuments in Khakassia, who also studied the Gate to the Salbyk Valley. The gate was examined by S. V. Kiselev and L. R. Kyzlasov in 1947, and then L. R. Kyzlasov published a drawing of it. In the 1950s–1990s, L. R. and I. L. Kyzlasov repeatedly studied various objects in the Salbyk Valley, studied objects from Salbyk (Kyzlasov 1985a, b, 1986, 2008; etc.). In 1977, archaeologist M. P. Gryaznov invited L. S. Marsadolov, then still a history student, to photograph the burial mounds of the Tagar culture on the Yenisei.

Complex Research Stage

In 1992, 1994, 1996, 1998, 2008, 2010, and 2015 in the Salbyk Valley, the Sayan-Altai Archaeological Expedition of the State Hermitage Museum conducted multi-aspect studies under the leadership of Marsadolov (1993, 1998, 2010, 2014), and Marsadolov and Paranina (2011a; b). At that time, Akulov and Pauls (2008) studied the planigraphy and architecture of the burial mound field objects, Tuguzhekova (2000) described the health practices associated with the megalithic monuments.

At this time, the Great Salbyk Kurgan was studied in more detail: the plan of the fence made of stone slabs and the layouts located outside the fence were taken again; each of the 90 stone slabs of the kurgan and the nearest quarry were carefully measured and geologically studied. To determine the age of the kurgan, wood samples were taken for radiocarbon and dendrochronological analysis, which, based on archaeological analogies, made it possible to date the Tagar kurgan with an accuracy of 20 years—670 BC (Marsadolov 2014). Ancient drawings, signs, and tamgas were copied again, which provided new extensive materials for their dating, as well as for identifying the semantics, connections, and social organization of Tagar society. For the first time, paleoastronomical, geophysical, and metrological works were carried out, a circular theodolite survey was conducted from its central point to determine the connection of the fence elements with geographical and astronomical objects on the horizon line. In 2010, a generalizing monograph on this mound was published (Marsadolov 2010).

In the 1990s, *the Salbyk Gate* was explored by teams from the Sayan-Altai Archaeological Expedition of the State Hermitage Museum—SAAE SH (L. S. Marsadolov), The Khakass Research Institute of Language and Literature (Ya. I. Sunchugashev, N. N. Kuznetsov) and The Institute of Archaeology of the Russian Academy of Sciences (I. L. Kyzlasov). The monument consists of two vertically installed stone slabs, oriented to the cardinal points with an accuracy of 1°–2°. The southern slab (male) is made of gray siliceous sandstone. It has a subrectangular shape with a high eastern edge (like most of the slabs of the Tagar burial mounds). The dimensions of the slab are 2.25 × 1.35 m, the thickness is about 0.45 m (in the ancient anthropometric system—5 × 3 × 1 cubits). On the narrow edge facing east, in the middle part, images-tamgas are carved. The northern slab (female) is hewn from reddish-brown sandstone, at the ends, it has a rounded shape. In 1997, the slab was installed vertically under the supervision of N. N. Kuznetsov (in the late 1940s it tilted to the south (Kuznetsov et al. 1996), then fell and lay on the ground for a long time). The height of the slab (before installation) is 3 m, the greatest width is 1.8 m (1 fathom, or 4 cubits), the thickness is up to 0.2 m in the lower and upper parts, and up to 0.45 m (1 cubit) in the middle.

Astroarchaeological studies by SAAE SH in the 1990s showed that the two slabs of the Salbyk Gate are oriented with their wide side faces to the north and south, and their narrow sides along the line east–west (Fig. 2a, b). This line corresponds to the sunrise and sunset on the days of the spring and autumn equinoxes (Fig. 2c). Kyzlasov (1986, p. 139) wrote that the "Big Gate" of the Valley of the Kings, like the "gate" of the fences of the large Salbyk burial mounds, "face the east, towards the rising sun" as an indication of the movement of the souls of the buried. This does not exclude the possibility of using the Salbyk Gate in ancient times to determine the days of the vernal equinox. According to ethnographic data, until recently, on the day of the vernal equinox on March 22, the Khakass people left their aal for the nearest hill and greeted the sunrise, the beginning of the New Year ("Chyl Pazy" = "The Head of the year") (Marsadolov 2010).

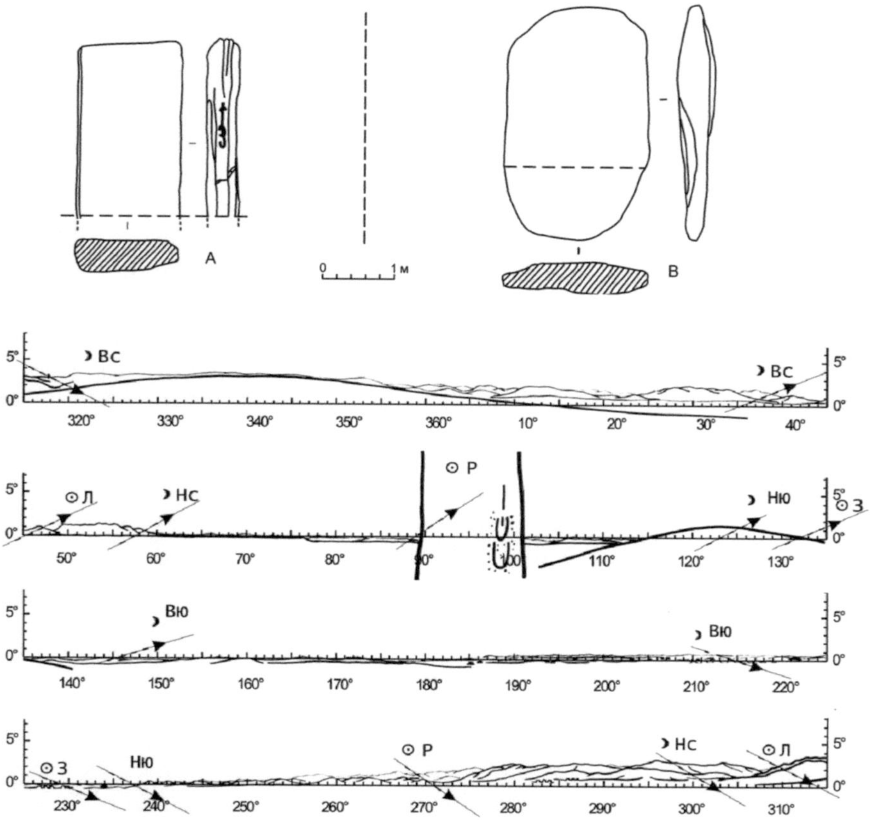

Fig. 2 Gates of the Salbyk valley. **a, b** views and sections of stone slabs: **a** southern slab; **b** northern slab. **c** Circular panorama from the center of the object. The horizontal axis shows azimuths in degrees (converted from astronomical north), and the vertical axis shows the height above the horizon. Legend: arrows pointing up (sunrise), down (sunset) of the Sun and Moon on astronomically important days. The Sun: З—winter solstice; Л—summer solstice; Р—equinox (spring and autumn). The Moon: С—northern, Ю—southern, В—high, Н—low. Based on materials from the expedition of L. S. Marsadolov, astronomical calculations by V. L. Gorshkov

Current Stage of Research of Objects of Salbyk Valley: Methods of Geospatial Analysis

Clarification of Spatial Characteristics

Earlier, we addressed the problem of inaccurate orientation of heritage sites on plans in archaeological reports and scientific publications, including those presented on the UNESCO World Heritage website (Paranina and Paranin 2022). An analysis of Russian and foreign scientific publications showed that determining the north in diagrams and plans can be difficult or impossible for several reasons: (1) the cardinal directions are not indicated; (2) the north is indicated with an error; (3) the original high-quality materials are used incorrectly; (4) the magnetic north is indicated, but the year of topographic survey is not indicated; (5) the supplemented topographic plans are superimposed on the historically preceding topographic base. The revealed facts convince us of the need to increase the level of general geographical culture, interdisciplinary cooperation and special training in new methods of geospatial analysis. This is facilitated by the educational program "Spatial and digital analysis in geography", developed and implemented with the participation of the authors (Zolotova et al. 2017a, b; Sidorina et al. 2024).

To clarify the spatial characteristics of objects, it is convenient to use the method of superimposing multi-temporal and multi-faceted data. Shooting from a quadcopter allows you to get a detailed image of the object, as well as decipher the landscape and topographic situation around it. Comparison of the plan image with a satellite image provides a more accurate determination of the direction to the north than working with a compass (taking into account the fact that the slabs of the mound fence are not level—they have become crooked over time). Then these materials can be compared with topographic plans in the QGIS geoinformation program.

For example, combining the topographic plan and photographs of the Great Salbyk mound gave a good match of the contour (Fig. 3). At the same time, it can be seen that individual slabs on the plan are not marked very accurately, especially the "gate" of the mound, located in the eastern part.

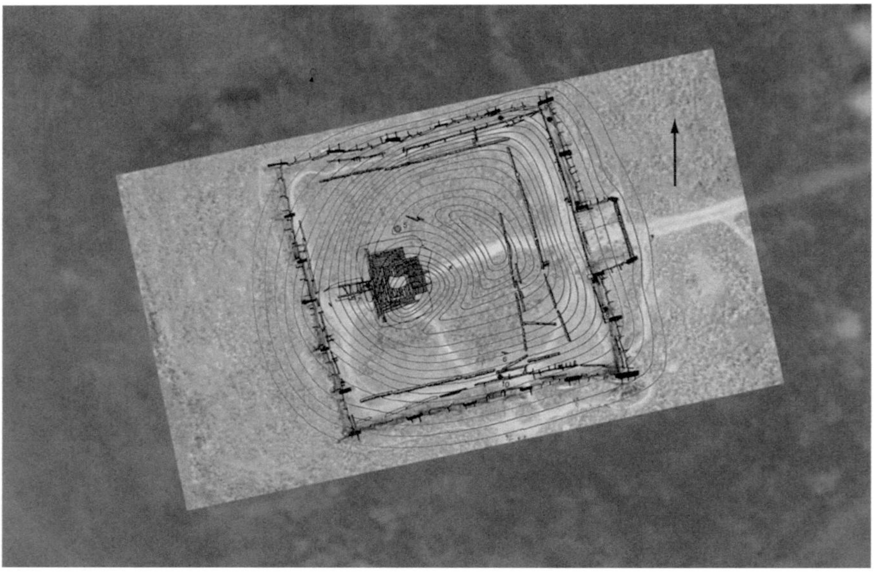

Fig. 3 Clarification of spatial characteristics of the Great Salbyk Kurgan (burial mound): a method of superimposing multi-temporal and multi-faceted data

Comparison of the images of the arrows indicating the direction to the north shows that the deviation on the topographic plan is 10.5° to the east of the geographic meridian, which significantly exceeds the magnetic declination of 3.4848° E (3°29′28″ E), determined using calculators (https://planetcalc.ru/320/). The identified discrepancies allow us to make the necessary corrections. This is important for the correct scientific description of The Great Salbyk Kurgan and for the accuracy of the archaeological and paleoastronomical studies that have been carried out at this site by archaeological specialists for several decades. Similar studies are useful for clarifying the orientation and planigraphy of other heritage sites.

Correlation of the Structure of Linear Objects with the Landscape Pattern and Azimuths of Astronomically Significant Points of the Horizon

Comparison of the directions of linear elements of cultural heritage sites and natural landscapes with the azimuths of the moon and sun rises/sets allows us to determine the rational basis for the sacralization of artificial and natural objects in the process of solving specific problems: identifying connections in a man-made spatial system; determining possible information (instrumental and navigational) functions of objects; reconstructing the level of astronomical knowledge and various forms of its expression in mythopoetic models of the world preserved in folk tradition.

Field studies of the spatial structure of the Salbyk Valley burial mound field began in 2010 (L. S. Marsadolov, G. N. Paranina, V. G. Borgoyakov, S. A. Ugdyzhekov, V. V. Tarakanov, S. O. Taskarakov). The distances between the extension slabs near the largest burial mounds were measured using GPS, a theodolite, and a tape measure (Marsadolov et al. 2010). Linear relief elements (mainly of erosional genesis) were identified using space images and a topographic map. Astronomically significant azimuths were calculated using the paleo-astrocalculator of the Central (Pulkovo) Astronomical Observatory of the Russian Academy of Sciences. Combined circular diagrams, constructed in the Microsoft Excel program based on a summary table of data, clearly showed areas of coincidence and divergence of natural and man-made elements of the spatial system (Fig. 4).

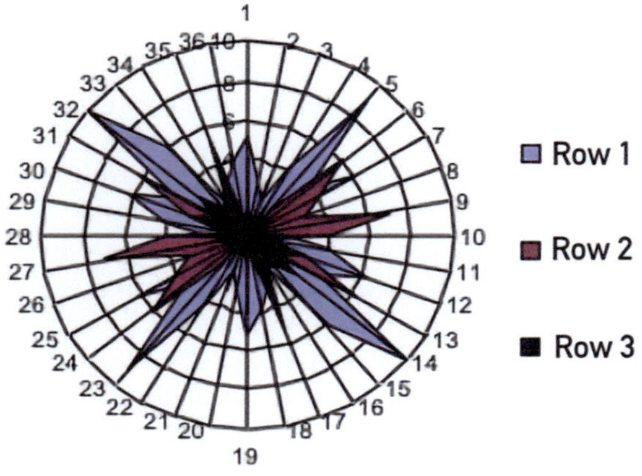

Fig. 4 Main directions of the landscape and objects of the Salbyk Valley. Number of linear elements of the landscape: Row 1—plain, Row 2—mountains; Row 3—number of directions measured between large burial mounds in 2010 (interval 10°, the geographic meridian corresponds to the line marked by the number 36)

High correlation of directions in planigraphy of archaeological object, structure of landscape and azimuths of key points of near-horizon astronomical observations became the basis of new approach to definition of the concept of «sacredness»: natural, natural-anthropogenic and anthropogenic objects are sacred, providing preservation, transfer, and use of vital information about the order of geographic space and time (Marsadolov and Paranina 2011a, b). Such approaches allows to assume that in ancient system of economic zoning of the territory the main criterion of choice of sacred landscape was its ability to perform information function.

Spatial analysis of objects of cultural heritage, located in Salbyk Valley and in other regions confirms the obtained results (Paranina Alina et al. 2022): (1) linear in form components of natural landscape—relief, hydrography, boundaries of large plant associations—can have convenient for observation and recording of astronomically significant azimuths; (2) directions needed for orientation in space and time can be distinguished by anthropogenic contributions of various natures—clearing of weathering products at the foot of stone remnants, giving sculptural forms according to the principle of "remove the excess", installation of stone blocks, notches, petroglyphs; (3) such objects often have their own legend, in some cases they are associated with a living tradition (folk holidays, bringing gifts, addressing prayers).

Geolocation of Objects and Reconstruction of Stages of Evolution of Archaeological Complexes

The diversity of ages of the elements of prehistoric heritage sites makes it possible to trace the stages of development of navigation technologies: (1) sacred landscape (tools are relief and natural outcrops of bedrock: micro-relief fixes a convenient observation point, the horizon line is a space–time scale, compass and calendar), (2) the stage of local networks (man adds artificial tools to the sacred landscape—megaliths, large blocks of dense rocks resistant to displacement and destruction, they fix the points of the horizon at which calendar astronomical events are observed (for example, paired objects work on the principle of "aim"—"fly sight"), (3) the stage of regional networks (determining the latitude by the shadow of the solar gnomon makes it possible to link local networks with each other), (4) the historical stage of abstract modeling (development of graphic signs and symbolic images, application of tamga-shaped signs, anthropomorphic and zoomorphic symbols to the surface of slabs drawings), (5) the current stage of new technologies of navigation, communication and modeling (geodetic network points have been installed on the tops of some mounds). Despite the diversity of forms, the geolocation of objects is preserved—their main value and characteristic (Paranina 2017; Paranina and Paranin 2017). Figure 5 shows the main stages of the development of the Salbyk mound field, reconstructed on the basis of identifying elements of different ages: Stage 1—the "fertility" stele and "extension" steles of the mounds (Bronze Age); Stage 2—large mounds; Stage 3—groups of small mounds. Elements of the spatial system: red dots—the most ancient stones; squares—large burial mounds nearby; unfilled circles—small burial mounds in axial directions. Structural relief lines: ridges, keel lines, edges, and back seams are drawn according to the method of prof. A. N. Lastochkin. Basis of the map schemes: topographic map of the Salbyk Valley, plan of the burial mound field—compiled by Kiselev (1956), supplemented by Marsadolov and Paranina (2011a).

Reconstructions of Instrumental Functions of Oriented Spatial Systems

The combination of a detailed plan photograph and a satellite image makes it possible to clarify the spatial structure of the monument (Fig. 6a, b): the location of the centers of two stones deviates slightly from the midday line, which could be the result of not very precise installation of the northern stone during restoration. For the coordinates 53.93°N, 90.83°E, using the calculators of the height of the Sun and the length of the shadow [https://planetcalc.ru/320/], the lengths of the midday shadow of the southern stone with a height of $h = 2.25$ m were determined for the days of the solstices (SS—summer and WS—winter) and equinoxes (AE—autumn and SE—spring) for 2024: SS June 20—1.32; WS December 22—10.03; AE September 22—3.09; SE March 20—3.10. The days on which the midday shadow of the southern (male) stone reaches the foot of the northern (female) stone, the distance between which is 3.6 m, are determined: 2.10–3.55; 3.10–3.61; 10.03–3.59 m (Fig. 6c).

The obtained result can be explained in different ways: (1) the approach of the midday shadow of the southern stone to the foot of the northern stone marks the boundaries of the coldest time of the year, which do not coincide with the days of the equinoxes by 10 days; (2) the midday shadow should reach the foot of the northern stone on the days of the equinoxes, being an additional signal of the onset of a new astronomical season, available for observation during the active time of the day. In the latter case, it can be assumed that during the restoration of the monument, the northern slab shifted not only in longitude but also in latitude. The 0.3–0.5 m shift of the northern stone could have gone unnoticed, since it is comparable with the thickness of the installed slab (0.2–0.45 m). In addition, the exact location of the installation could not have been preserved, since the stone fell, lay horizontally for a number of years, and the pit was filled with soil. It is also possible to imagine that a lying stone prevented digging a hole in the old place.

The diagram shows that all midday shadows in winter fall on the northern stone, which functions as a “mirror”—the association is reinforced by the rounded shape of its upper edge. Since the length of the day in the winter half of the year in Khakassia fluctuates between 12 and 7.5 h, the plane of the northern stone can act as a kind of clock face: first, the eastern edge of the shadow of the southern stone appears on its plane, then the shadow of the entire top of the slab, then only the western edge of the shadow is visible, then it goes beyond the edge of the dial. In summer, the top of the shadow of the southern stone is significantly shorter than the distance between the slabs, but you can use the shadow of your own figure, for this it is better to stand opposite the center of the slab: in the first half of the day the shadow will fall to the left, in the second—to the right of the imaginary axis, at midday it will lie on it.

The transverse hypsometric profile crossed the ridge. Two-rock Ridge at an azimuth of 40° and passed through the “Gate”. The direction of the ridge is about 130°/310° close to the sunrise at winter / sunset at summer solstice. Such geometric properties of the linear elements of the relief could contribute to the choice of the place for installing the astronomical instrument and were reflected in the status of

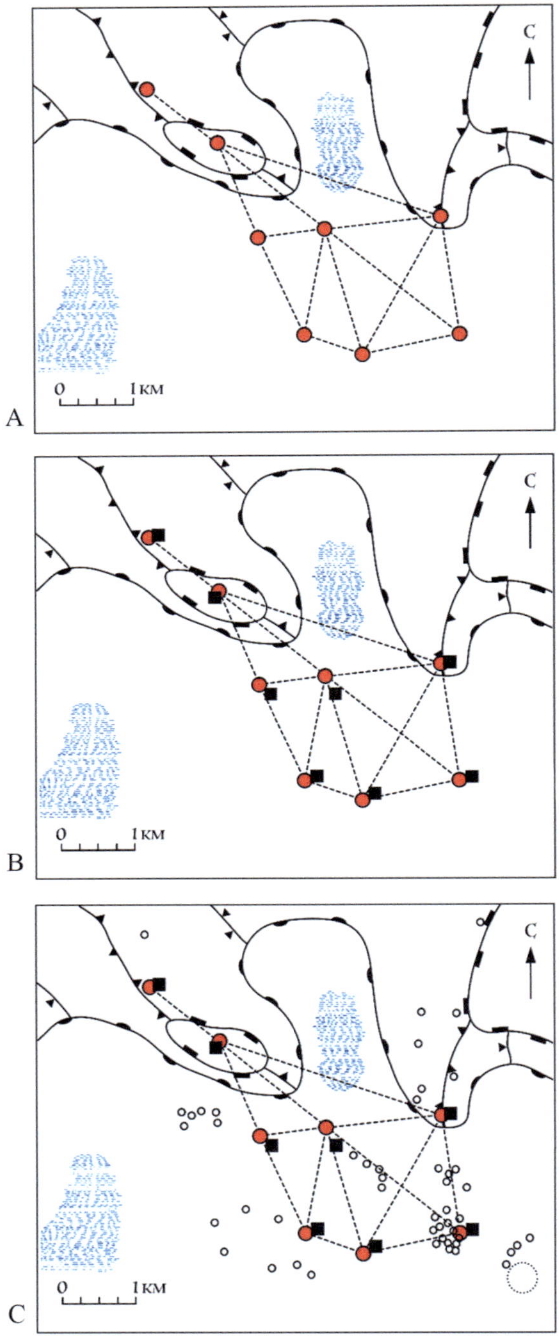

Fig. 5 Stages of development of the Salbyk burial mound field. **a** Stage 1: "fertility" stele and "extension" steles of burial mounds (Bronze Age); **b** Stage 2: large burial mounds (Tagar time); **c** Stage 3: groups of small burial mounds. The scheme was compiled by R. V. Paranin

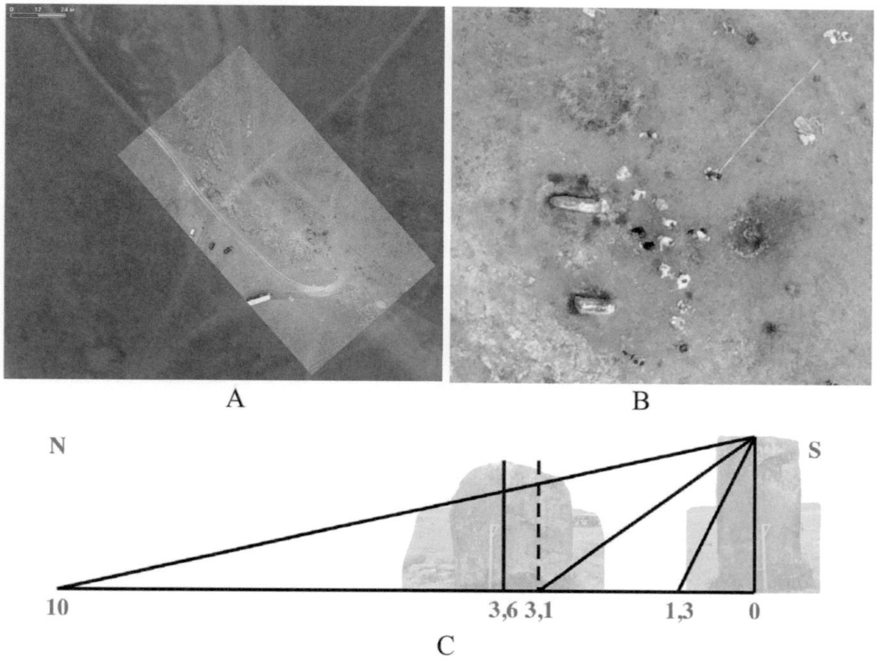

Fig. 6 Two stones as a sundial-calendar: **a**, **b** plan view, **c** diagram of the position of the midday shadow from the southern stone on the days of the equinoxes and solstices: dotted line—possible position of the northern stone before restoration. Materials of R. V. Paranin and A. N. Paranina. The scheme was compiled by R. V. Paranin

the sacred landscape—a natural territory that preserves and reproduces vital standards of space and time. It can be assumed that the depression in the relief is of artificial origin, and one of its functions in ancient times was to ensure movement in the meridional direction (across the steppe along the mountains). At the same time, the landscape is the first astronomical instrument that records the observation point and the direction to astronomically significant points of the horizon. The modeling of the physical surface and the subsequent installation of sighting devices served the development of the calendar folk tradition and formed the technological and cultural multi-layered nature of the sacred monument.

General Geographic Education and the Effectiveness of GIS

Examples of geospatial analysis of cultural heritage sites prove that teaching GIS methods is necessary for improving not only geographical but also humanitarian education (Andreeva and Litvinova 2015). However, software tools will not be effective without basic geographical knowledge. This social demand can be met in different ways: (1) involving professional geographers in the implementation of humanitarian projects; (2) including geographical disciplines (general physical geography and cartography) in the curricula of all humanitarian specialties; (3) developing continuous geographical education, which is carried out as a hobby throughout life, including in tourism (Zolotova et al. 2017a, b; Andreeva et al. 2023).

The cognitive activities of tourists at cultural heritage sites associated with navigation technologies (Paranina 2011, 2017; Paranina and Paranin 2017; Paranina Alina et al. 2022) can be organized in accordance with the technological level of ancient tools (Table 1). To the traditional photographic recording of objects for tourism are added: (1) observations of luminaries (fixing directions with a compass); (2) calculations (using astro-calculators); (3) drawing and analyzing graphs, topographic plans. Based on regular visits to objects, it is possible to create a database (personal or general, updated on the sites of the objects). Analysis of the results in the form of map diagrams and comparison with the model of the celestial sphere motivates self-education and contributes to the improvement of general geographical culture. Conclusions, presented in the form of presentations and video clips, competitions and conferences can become forms of discussion and, more effectively than advertising, promotion of a tourist product to the masses. In all cases, such activities provide active recreation associated with a long stay in the open-air.

Interaction with the surrounding world, based on careful observation and solving cognitive problems, improves positive personality traits: trains thinking, attention, and memory, fosters a respectful attitude towards nature and ancient man, towards natural beliefs and folk traditions that have come down from the depths of centuries. In turn, this contributes to increasing the viability of modern society: the development

Table 1 Types of cognitive activities of tourists at prehistoric sites of primary navigational purpose

Technology activity	Landscape-tool	Natural man-made tool	Man-made tool
Photographic recording (from a central point)	Panoramic photography and natural objects	Planned filming and natural and man-made objects	Filming of the tool and its parts
Stargazing (fixing directions with a compass)	Crossing the horizon: connection with the horizon shape and linear elements of the landscape	Crossing the horizon: connection with landscape elements and Archaeological complex	The movement of a shadow or a beam of light across an artificially arranged area reveals the functions of the elements
Calculations (using astro-calculators). Drawing and analysis of graphs, topographic plans	Sunrise/sunset azimuths	Sunrise/sunset azimuths	Length and direction of shadow
Results	Database creation. Spatial analysis. Presentation and discussion of the results obtained		

of systemic, spatial, critical, creative, cosmic, and ecological thinking; strengthens interest in the diversity of adaptations in nature and culture, increases social tolerance and respect for ancestors, mutual respect, and self-esteem.

Conclusions

Modern methods of geospatial analysis allow us to specify the geolocation and orientation of cultural heritage sites, as well as to deepen our understanding of the functions of ancient objects. Using the example of the objects of the Salbyk Valley in the Republic of Khakassia (Southern Siberia), we can see that sacred objects performed a number of important social functions: ritual and memorial (with the addition of new objects), astronomical and navigational (the linear arrangement of objects corresponds to astronomical calendar azimuths) and the function of social communications (collective ritual activity and calendar holidays). The criterion for choosing a sacred landscape could be the coincidence of linear relief forms with calendar astronomical azimuths and a distinct broken shape of the physical horizon, convenient for use as a time scale. Comprehensive field practices and the completion of qualifying work allow students to be involved in real scientific research. Geographical students already in their first year in the discipline "Cartography with the basics of topography" acquire skills in using geodetic instruments and master the QGIS software product. In the course "Geoinformatics" students improve their skills in working with geoinformation systems, working on databases for the objects of their research.

The tasks of research and preservation of the cultural heritage of mankind can be solved based on modern GIS methods. An indispensable condition for the effective use of new software tools is the development of general and continuous geographical education.

References

Akulov AG, Pauls ED (2008) On the study of Tagar burial mound architecture. In: Proceedings of the II (XVIII) all-Russian archaeological congress in Suzdal, vol II. Moscow, pp 5–8

Andreeva TA, Litvinova MV (2015) Cartographic and geoinformation research work of students. In: Lazebnik OA (ed) The collection: geodesy, cartography, geoinformatics and cadastres: from idea to implementation. Collection of materials of the international scientific and practical conference, pp 314–319

Andreeva TA, Pozdnyakova NA, Shtykova NB, Obraztsov PA, Stroganov YV (2023) History of the emergence and development of maps for tourists. In: The collection: natural and cultural heritage—interdisciplinary research, preservation and development collection of scientific articles based on the materials of the XI all-Russian scientific and practical conference with international participation. St. Petersburg, pp 290–296

Devlet MA (1976) Bolshoy Salbyksky Kurgan: the grave of a tribal leader. From the history of Siberia. Tomsk 21:146–154

https://planetcalc.ru/320/

https://www.google.ru/intl/ru/earth/
https://www.livelib.ru/author/347337/top-a-n-lastochkin?ysclid=m1pktohj1v795422942
Kiselev SV (1951) Ancient history of Southern Siberia. Moscow, p 642
Kiselev SV (1956) Research of the Great Salbyk Kurgan in 1954 and 1955. In: Theses of reports at the session of the department of historical sciences and the plenum of the IIMK, dedicated to the results of archaeological research in 1955. Moscow, pp 56–58
Kuznetsov NN, Marsadolov LS, Sigedina NG (1996) Stone "gates" from Salbyk. In: Priesthood and shamanism in the Scythian era. Materials of the international conference. St. Petersburg, pp 181–187
Kyzlasov IL (1985a) Pyramids of the Salbyk Valley (guide). Abakan
Kyzlasov LR (1985b) Tagar hand mill and its significance: in memory of the teacher. Soviet Archeol 3:65–71
Kyzlasov LR (1986) Ancient Khakassia, p 295
Kyzlasov L (2008) Pyramids of the steppe kings: treasures of the culture of Khakassia. In: Heritage of the peoples of the Russian Federation, pp 382–385
Marsadolov LS (1993) Research of the Sayano-Altai archaeological expedition in 1992. In: Report archaeological session of the State Hermitage Museum. Brief abstracts of reports. May 1993. St. Petersburg, pp 3–5
Marsadolov LS (1998) Barrows in Salbyk: Siberian Giza. In: International conference on primitive art. Abstracts of reports, pp 172–173
Marsadolov LS (2010) The Great Salbyk Kurgan in Khakassia. Abakan, p 128
Marsadolov LS (2014) Dates of the Great Salbyk Kurgan and its stage in Khakassia. In: Proceedings of the IV (XX) all-Russian archaeological congress in Kazan, vol II. Kazan, pp 211–217
Marsadolov LS, Paranina GN (2011a) Salbyk archaeological complex as an object of natural and cultural heritage. Bull Russian Geogr Soc 143(2):79–90
Marsadolov LS, Paranina GN (2011b) The first comprehensive studies of megalithic objects in the North-West of Russia: methodology and techniques. In: Geography: problems of science and education. LXIV Herzen readings. Proceedings of the annual international scientific and practical conference LXIV Herzen readings, dedicated to the memory of A.M. Alpatyev, St. Petersburg, Herzen State Pedagogical Univ., April 21–23, 2011, pp 45–52
Marsadolov LS, Paranina GN, Ugdyzhekov SA, Borgoyakov VG (2010) Information analysis of the sacred space of the Salbyk Valley. In: Modern transformational economic and socio-political processes. Abstracts of reports of the II international scientific-practical conference (Abakan, October 27–29, 2010). Abakan, pp 132–138
Marsadolov LS, Paranina AN, Paranin RV (2023) 283rd anniversary of monitoring the state of The Great Salbyk Kurgan in Khakassia, LXXVI Herzen readings. In: Geography: development of science and education—proceedings of the international scientific-practical conference April 19–21, vol I. Herzen State Pedagogical Univ., pp 42–54
Miller GF (1937) History of Siberia, vol I. Moscow-Leningrad, p 607
Paranina GN (2011) Navigation in space, time and information: the basis of life support. In: Ecology of ancient and traditional societies. Institute of Northern Development Problems SB RAS, Tyumen, pp 115–119
Paranina A (2017) Archaeological objects as elements informational life support system and as sources of information about evolution of environment. Geomate 35:100–107
Paranina A, Paranin R (2017) Information in geographical space as the basis of crossdisciplinary researches in culture geography. Eur J Geogr 8(3):67–77
Paranina A, Paranin R (2022) On the determination of the geographic north on archeological plans in connection with the problem of the quality of geographic education. Athens J Sci 10:1–11
Paranina Alina N, Marsadolov Leonid S, Grigoriev Alexey A (2022) Geographical tools of sacred navigation in the historical development of Siberia. In: Humans in the Siberian landscapes. Ethnocultural dynamics and interaction with nature and space. Cham, pp 33–58. https://doi.org/10.1007/978-3-030-90061-8_2

Sidorina IE, Syuzyumov AA, Rakova AI, Andreeva TA, Artemyeva OV (2024) Geoinformation mapping in modern ethnographic, confessional and ethnoconfessional studies. Kunstkamera 2(24):61–79

Sunchugashev YI (1989) Monuments of irrigated agriculture in ancient Khakassia. Krasnoyarsk 12:104

Tuguzhekova VN (2000) Salbyk mound. Abakan literary, Abakan, pp 71–72

Zolotova TI, Andreeva TA, Litvinova MV (2017a) Mapping of cultural and spiritual heritage sites for tourism purposes. In: The collection: geodesy, cartography, geoinformatics and cadastres. From idea to implementation. Collection of materials of the II international scientific and practical conference. St. Petersburg Association of Geodesy and Cartography, pp 270–276

Zolotova TI, Andreeva TA, Litvinova MV (2017b) Mapping of cultural and historical heritage: experience of the department of cartography and geoinformatics of St. Petersburg University. Quest Geogr 144:292–305

The Game Theory as a Collaborative Tool for Managing Geographic Information Systems (GIS) Applied to Cultural Heritage

Salvador Castillo-Rivera and Olaia Fontal Merillas

Abstract Nowadays, managing data is essential to making knowledgeable decisions. Geographic Information Systems (GIS) have developed into a tool for yielding fitting analytical visualizations. These allow us to discover variables, patterns, distributions, etc. The database of a GIS is often composed of two principal components: graphic information and tabular data. The first includes information concerning the location of characteristics, and the second holds non-spatial information related to the features. A GIS is developed to collect, run, and derive an extensive diversity of geographically referenced data applicable for investigation, management decisions, etc. Nevertheless, obtaining information supposes the availability of funding in some cases. The economic limitations may reduce the options to fund a detailed search to get accurate information. The researchers and educators in heritage are at a point wherein they will obtain considerable advantages from GIS usage if the collaboration is considered between university students, heritage projects, research groups, etc. This technology lays out advantages that have not been entirely developed in heritage education. So, they should learn to manage cooperation and decision-making in heritage education. Game theory can be used to manage the cooperation and the flow of information between stakeholders in the setting up of databases in a heritage context. This theory can provide a framework to address the distinct situations that may derive from this setting. In this chapter, we develop an approach based on game theory methodology to help focus on and analyze real-life cases in cultural heritage environments, enabling decision-making once the context has been analyzed.

Keyword Geographic information systems · Cultural heritage · Game theory · Collaboration

S. Castillo-Rivera
Department of Mathematics, Universidad Carlos III de Madrid, Leganés, Madrid, Spain

O. F. Merillas (✉)
Didáctica de la expresión musical, plástica y corporal, Universidad de Valladolid, Campus Miguel Delibes, Valladolid, Spain
e-mail: olaia.fontal@uva.es

R. de Miguel González and P. Rivero Gracia (eds.), *Geospatial Technologies for Heritage Education*, Key Challenges in Geography,
https://doi.org/10.1007/978-3-031-96372-8_3

Introduction

Nowadays, managing data is essential to making knowledgeable decisions. Geographic Information Systems (GIS) have developed into a tool for yielding fitting analytical visualizations. These allow us to discover variables, patterns, distributions, etc. This is carried out through maps linked to data. Liu et al. (2024) have indicated that GIS are progressively significant in the domain of cultural heritage conservation. This promotes the setting up of dynamic information management systems and operates as powerful platform for research and visualization. This investigation area is undergoing a phase of steady development. The study recommended the incorporation of digital technologies into the humanities, highlighting the demand for a heritage database provided with standardized data exchange protocols to sustain visualization and analytical functions.

GIS technology combines spatial and attribute data of heritage sites, which is a robust tool for better compression, preservation, and management of heritage resources. Huang (2024) revised state-of-the-art in GIS applications between 1994 and 2023 from the Web of Science database. He has put forward proposals in the field of GIS-Heritage, such as concentrating on data integration and technological innovation. All of these should be carried out with an interdisciplinary and international collaboration. Basir et al. (2014) have pointed out that heritage buildings are subject to threats from natural catastrophes, uncertain weather, pollution, etc. To maintain this, the recording and documenting of heritage buildings is needed the three-dimensional (3D) information has a significant function in this topic. Combining 3D modelling with the technology of GIS will enable to generation of a database that can drive different analyses of spatial data in the shape of a 3D model. The outcome of data acquisition will evolve a guideline for 3D model development. The model will be sent to the GIS format to create a database for heritage building conservation. Bruno et al. (2020) have reported that the adequate integration of BIM (Building Information Modelling) and GIS in a unique information system is an available research area, particularly when tackling Cultural Heritage. They described research on the evolution of a web information system capable of integrating BIM and GIS data, with a special emphasis on the study of the historicized city and its main buildings over time.

An essential tool in educating and researching fields is the data and management software associated with urban planning, geomatics, etc. The interchange of data between information systems is a challenge. However, its solution may assume huge advantages to those who operate with them. Malinverni et al. (2020) have noted that information management systems lead to modifications to the methods of data production, documentation, and cataloging in the building data processing. The integration of information systems means an efficient solution for determining suitable projects, such as conservation and restoration processes, providing opportunities to integrate heterogeneous data. So, they have developed a model, compared with the standard GIS one and have shown that it can be achievable to interoperate modified data towards a further unified data model without losing any information.

Martínez-Hernández et al. (2023) have set out that GIS holds an extensive presence in the present geographic research. They state that the new academic curricula commence to contemplate this fact. Nevertheless, there is a considerable deficiency in the teaching of geodigital skills, required to didactically employ GIS in teacher activity. To encourage these skills for didactical purposes, they fixed a theme with a geographical approach as urban heritage. The aim was to develop a workshop focused on digital mapping via a story maps application using web-GIS technology. Taking into consideration a framework of competencies supplied by the specialist literature, the workshop tasks were associated with geographic reasoning skills together with geodigital competencies, heritage interpretation capabilities, etc. The workshop was focused on students registered in a higher education teacher training program. Pepe et al. (2021) have implemented a Scan to BIM method applied to the cultural heritage field to set up a 3D GIS model which can link various databases to enable a multidisciplinary practice. The procedure was used in the case study of a rock church.

This chapter presents an approach to derive a more suitable understanding of decision-making by using game theory in an interdisciplinary collaboration environment that facilitates the interchange of information and data of GIS in a cultural heritage setting. The results of this work shed light on the management and the cooperation between them.

The outline of the manuscript is as follows: the materials and methods section puts forward a description of GIS databases and game theory. The following section presents the outcomes of the work applying the stag hunt game between different investigation teams in a cultural heritage environment using GIS technology. Finally, the discussion is presented in the last section.

State of the Matter

GIS connects numerical data to cartographic sheets, which makes this technology an accurate tool for exploring and visualizing data from a territorial standpoint. This certainly improves decision-making when the geography zone is essential for analyzing their features with different study purposes. These databases enable more complete information to be accessible for decision-making. The data collected can be utilized by a large number of users and the cooperation between them can relieve the economic cost and the cost of time that can be involved in the access and the building of a certain database. Due to the variety of complex components that exist as well as the different interests that can come together, such as physical, human, cultural, and artistic, that can be derived from the usage of geospatial technologies.

Nishanbaev et al. (2021) have indicated that the cultural heritage field has begun executing connected open data concepts on cultural heritage data. They have presented a cloud architecture that tries to improve digital cultural heritage exploration by integrating 3D digital cultural heritage models with connected open data from DBpedia (project to extract structured content from the information produced in the Wikipedia project) and GeoNames (user-editable geographical database unrestricted via different web services, under a Creative Commons Attribution license) platforms using web-GIS technologies. The model was validated with particular case studies, looking for expert feedback. Santos et al. (2021) have pointed out that in architecture GIS have been restricted to regional and urban development applications. However, they have claimed that its usage has made progress in handling urban planning and architectural heritage management subjects. Digital technologies have extended the options for developing more interdisciplinary projects. A more adequate integration of GIS is fundamental in architectural design, heritage, etc. Most architecture programmes utilize GIS technology but still depend on the student's readiness to learn GIS basics. Thereby, there exists a necessity for educational training to leverage the entire potential of information analysis and decision support that GIS permits. Students should explore the usefulness of GIS tools to examine case studies and to satisfactorily understand how GIS technology can help the process of planning and management. Forthcoming developments must cover earlier and broader teaching of GIS, tools, and techniques. This will promote the development of postgraduate studies and research projects such as 3D GIS and BIM, among others. It should involve the right transfer of knowledge to the teaching community.

Matrone et al. (2019) described the various attempts and outcomes obtained from the integration of a Historic Building Information Modelling (HBIM) model into an organized spatial database and its 3D display in a GIS project. This study was linked to the European ResCult (Increasing Resilience of Cultural Heritage) project, wherein a database for multiscale analyses was determined. The trial had as purpose of understanding how the data and attributes put in the HBIM are restored and managed when addressing a GIS database, moving from the Industry Foundation Classes (IFC) to the Geography Markup Language (CityGML) standard, via the Feature Manipulation (FME) software. Matrone et al. (2023) have set out the findings of the Main10ance project. It focused on generating a combined Historic Building Information Modelling–GIS (HBIM–GIS) platform to hold-up the maintenance and conservation programs for the historic built heritage. For a case study, set up a unified system that semantically and spatially described the architectural elements, the surroundings, and the indoor scenarios. The database allowed interoperability amid diverse actors and domains, collecting heritage-related information essential for restoration and conservation objectives. Yagoub and Yammahi (2022) have presented research to evaluate the spatial distribution of natural risks and their closeness to heritage sites in the United Arab Emirates by applying remote sensing and GIS methods. The database, methodology, and outcome of this study may be helpful for many organizations, such as culture, heritage, tourism, urban planning, etc. The compilation of a GIS database focusing on the heritage sites in the United Arab Emirates has been a relevant stage for forthcoming studies associated with the sites. The database is unrestricted for interested users upon demand. Arrighi et al. (2023) have done a study to develop a GIS-based Flood Damage Index for Cultural Heritage (FDICH), which is planned for large-scale applications, such as regional to determine damage hotspots and help hazard mapping and management. The findings enabled for determination of major cultural heritage damage hotspots and the calculation of

FDICH displayed the potential for cross-compare damages to the Po River District which is the largest in Italy.

Planned maintenance implies a strategy to promote the conservation of architectural heritage to control invasive restoration activities. Colucci et al. (2022) have proposed the management of a maintenance program via Building Information Modelling (BIM) and GIS domains. A unique spatial database was set up and split into different Levels of Detail, respectful of geographical standards. The database is user-friendly for the specialists interested. Dhonju et al. (2018) have shown an online geo-crowdsourcing system called Share Our Cultural Heritage (SOCH). It can operate for large-scale heritage documentation and sharing. This can be helped by web and mobile GIS; cultural heritage data can be obtained through transportable devices. The data are georeferenced and shown via web mapping.

The planned maintenance of cultural heritage is currently supported by innovations in the digital domain, more specifically by digital models. Main10ance is an INTERREG project that seeks to execute a plan for the maintenance and conservation of the historical cultural heritage. The project involves the case study is the system of the Sacri Monti of northern Italy and Switzerland. The strategy phases are split into: study of the historical architectural heritage current. Data processing and development of 3D models with the support of BIM software; incorporation of the identical in the geographical setting via GIS aid; design of a database which yields interoperability between the different parts and that gathers data on the features of the heritage for maintenance and conservation objectives (Bonfanti et al., 2021). Rechichi (2020) presented Chimera, which is a system particularly, designed for the domain of Cultural Heritage. It is united to the reduced set of ready-to-work data systems that enable the together use of diverse integrated data, for instance, 3D models, 2D maps, images, etc. Further functionalities are also presented that permit it to operate at a multiscale approach, which includes changing from a BIM-style system to a GIS-style information system. This provides the opportunity to handle transversal information and to study each case from diverse insights and at distinct levels of detail.

Material and Methods

GIS connects numerical data to cartographic sheets, which makes this technology an accurate tool for exploring and visualizing data from a territorial standpoint. This certainly improves decision-making when the geography zone is essential for analyzing their features with different study purposes. The database of a GIS is often composed of two principal components: graphic information and tabular data. The first includes information concerning the location of characteristics, and the second holds non-spatial information related to the features (Aschenbach, 1991). GIS is a tool to capture, accumulate, update, operate, examine and display all records of geographically referenced data. GIS can be a tool for interdisciplinary research because it incorporates diverse data sources, allowing to help researchers across disciplines (Rickles et al., 2017).

As stated by Spiridon et al. (2016), the methodology of scientific research in conservation science needs a multidisciplinary approach that demands artistic, scientific, and technological skills, etc. This means the work teams have extended and have become gradually more complex. Thus, in the scientific research of the conservation state of cultural heritage new technical and methodological findings, such as GIS are used. These systems enable the geographic and cartographic representation of the features contained in the relational databases. For instance, these types of bases allow the visual representation of the geographical distribution of historical monuments. GIS are practical tool in the scientific procedure of managing and conserving cultural heritage. They provide the option to gather, review, and process complex data. The characteristics incorporated in the databases can be updated in real-time, being a meaningful element in monitoring, protection and valorization strategies of cultural heritage. Burrough and McDonnell (1998) put forward several factors that involve spatial data, which are fundamental when selecting data models and database approaches for any determined application; (a) to consider whether the real-world case of study is simple or complex. (b) If the kinds of entities employed to explain the situation are detailed or generalized. (c) To define if the data type utilized to create characteristics are nominal, topological, etc. (d) the items in the database describe entities that can be represented exactly or these are objects complex. In addition, it should be regarded whether their properties are exact, deterministic or stochastic. (e) To assess if the database objects describe discrete physical items or continuous domains. (f) To view if the database items are obtained by entire enumeration or sampling. (g) To examine if the database will be run for descriptive, administrative or analytical objectives. (h) To analyze if logical, empirical or process-based models are needed to get new information from the database and create inferences concerning the real world. (i) Finally, to assess if the procedure is static or dynamic.

Game theory examines the interaction between a set of agents. It is considered that individuals are rational, i.e., they pursue to maximize their payoff function. Every one of them knows this information and every individual acts strategically. If a player determines the actions without taking into consideration a reasonably selected payoff function, then the behaviour is not strategic. The main elements of a game are the players, who are the set of agents, such as individuals, corporations, etc., that interact in the game. The strategy is an entire plan explaining which actions a player selects in each feasible situation that the individual copes along the game. A game needs to list the payoff that every player receives under each feasible strategy (Espinola-Arredondo and Muñoz-Garcia 2023).

The stag hunt game describes a trade-off between playing it safe and depending on others to reach a more substantial gain. Two hunters, who can be called players one and two, can each select to hunt a stag, which supplies a relatively large or hunt a hare but considerably less filling. Hunting stags is difficult and needs cooperation. If either chases a stag independently, the probability of success is insignificant, while hunting hares is an individualistic action that is not accomplished in pairs. Therefore, hunting stags is most advantageous for society but demands "trust" between the hunters, and they have to unite forces between them (Tadelis 2013). So, the optimal strategy is to match the decision of the competitor. As stated by game theory principles, if one individual departs from the position, the other should also. If one individual determines to remain, the other should remain likewise. In the stag hunt game, no player holds exactly dominant strategies. When player two sets stag, player one is better off selecting stag also; whereas when player two chooses hare, player one is better off selecting hare (Espinola-Arredondo and Muñoz-Garcia, 2023).

Game Theory as a Tool for Managing GIS in Cultural Heritage Settings

As previously stated, the information management systems guide changes to the processes of data production, documentation and cataloguing in the building data processing (Malinverni et al. 2020). Digital technologies have expanded the opportunities for developing more interdisciplinary projects. A more satisfactory integration of GIS is basic in architectural design, heritage, etc. It is a requirement that the educational community forces the whole potential of information analysis and decision support that GIS enables mainly the students for their corresponding academic training (Santos et al. 2021). There exists a substantial lack in the teaching of geodigital skills needed to didactically use GIS in teacher activity (Martínez-Hernández et al. 2023). The incorporation of digital technologies into the humanities underlines the need for a heritage database supplied with standardized data exchange protocols to maintain visualization and analytical functions (Liu et al. 2024). All these tasks should be implemented with interdisciplinary and international cooperation (Huang 2024). So, the exchange of data between information systems becomes a challenge. The proper integration of information systems is an open investigation domain, especially in Cultural Heritage (Bruno et al. 2020). This can be capable of connecting different databases to promote a multidisciplinary practice. The database, process, and result of these guidelines can aid multiple institutions, such as culture, heritage, etc. This can supply the chance to drive transversal information from various understandings and at distinguishing levels of detail. Thereby, the students should realize that the procedures of database and GIS in heritage are associated with a diversity of components due to the complexity of human, physical, cultural and artistic characteristics implicated.

In multiple issues, designing and managing the GIS databases requires multidisciplinary teams that can cooperate between them. These bases aid in enhancing understanding of the heritage, significantly, for students who are preparing to become geography, history and art history teachers. Training in geospatial technologies will enable the corresponding stakeholders to establish their effectiveness and inquire into departmental and university procedures in time and real circumstances.

The interdisciplinary activity is promoted due to the possibility of sharing graphics and alphanumeric data (Pedelì 2013). Interdisciplinary investigation with GIS may be advantageous because researchers from diverse disciplines have the chance to generate something innovative. Nevertheless, GIS are difficult to learn and operate. To satisfactorily help interdisciplinary investigation with GIS, it is required to understand why researchers with interdisciplinary experience desired to operate it and how they learned it. So, the purpose should be to advise geography educators on establishing learning resources that may complete the existing learning methods employed by interdisciplinary researchers to enhance the learning knowledge of GIS. Interdisciplinary researchers make use of GIS to capture, analyze and visualize information. They mainly operate informal learning processes such as asking a more experienced professional, doing internet searches, GIS platforms, etc. The potential of GIS for interdisciplinary research has led to its incorporation in many projects. However, there exists a learning curve that must be overcome. The challenges related to the knowledge gap on GIS should be handled by enabling faster and more manageable uptake. GIS educators should help researchers in the performance of the application of GIS on projects to reach interdisciplinary research objectives (Rickles et al. 2017).

Collaboration between geography and further fields takes place often. Geography allows some use of its tools and methods to other topics and can incorporate them from outside. GIS are a fundamental tool for researchers. To start to work with GIS, students and researchers have to learn concepts of geographic information science, which is required to understand the internal language of GIS. Hence, it is needed for geographic knowledge and GIS concepts to combine spatial data and analyses of issues across fields. Interdisciplinary research facilitates working collaboratively across disciplinary boundaries. This scenario promotes the mixture of different procedures from the domains to attempt to answer multifaceted, real-world situations. The learning of GIS can be cumbersome for different causes, which can involve a lack of training hardware and software issues.

The use of GIS via interdisciplinary practices lays out chances to contribute to significant challenges between various disciplines because of determining spatial information at different scales. It is fundamental for managing issues such as biodiversity, cultural heritage, etc. As a case, we consider a geography interdisciplinary research team in a university department which is composed of senior researchers and students. Furthermore, it is also considered an art history research team in another department at the same university, which is composed of researchers and students. The first team present a large background in GIS, mainly, in data manipulation and data modelling. The second team is a specialist in the conservation process of religious heritage. The university has acquired the duty of developing several maps that emphasize the old religious city centre. The database should be updated in real-time, allowing the state institutions to determine the conservation policies more correctly, compared with the conventional methods. The local authorities have promised further grants and funding to the university whether the database is a tool that helps to enhance the cultural heritage successfully.

In this scenario, it is studied the respective incentives to teach GIS technology and learn it. The geography interdisciplinary research team could enhance their knowledge of manipulations of spatial and spatiotemporal data as well as the representation

		Interdisciplinary Team (IT)	
		Do not Teach GIS Technologies	Teach GIS Technologies
Art History Team (AHT)	Do not Willing To Learn	IT Uninvolved AHT Uninvolved	IT Loss AHT Small Win
	Willing To Learn	IT Small Win AHT Loss	IT Win AHT Win

Fig. 1 It shows the payoff outline for the stag hunt game for teaching GIS technologies between distinct research teams in a cultural heritage environment. The optimal condition is in the bottom right quadrant; the equilibrium is allocated in the top left quadrant. There are two situations in which one team does not trust the other

of regularized spatial and spatiotemporal reality through data models in some particular settings that they have not already handled (Hunt the Hare) and not taking the time to teach GIS technologies to the students and researchers of the conservation team. Both departments need additional funding for their corresponding projects; the interdisciplinary team must teach procedures in GIS technologies to the art history team to improve the monitoring, protection and preservation procedures of local religious heritage (Hunt the Stag). This involvement is advantageous to the geography department in the long run. Nevertheless, if this department spends time teaching, they expect the other department to be prepared to manage cases that can help the development of the database and the implementation of several maps.

From the art history team's insight, not being implicated in this dynamic can be possible, and they can reserve time for other tasks in the department (Hunting the Hare). If students and researchers learn from the geography department GIS technologies, they will address some issues with independence. Hence, the work that the interdisciplinary team has to do is diminished. As a result, both departments take advantage (Hunt the Stag), see Fig. 1.

Conclusions

Game theory tackles decisions in which the outcomes concern the actions of two or more players. Each one of them holds two or more manners of acting, they are called strategies. This theory is the science of strategy and decision-making. A game is a theoretical framework with a specified group of rules. The players act rationally and in their self-interest to improve their objectives. The outcomes can be considered as the effect of a conflict between decisions taken for self-interest and reciprocal benefit. The game theory states a means of outlining the fundamental structure of interaction and its representation in terms of a strategic game. This chapter has set out a conceptual tool to manage GIS applied to cultural heritage.

The GIS are developed to collect, run, and derive an extensive diversity of geographically referenced data applicable for investigation, management decisions, etc. The databases associated with these technologies are the guide of this study because they possess a remarkable role in this setting, and they enable more complete information to be accessible for decision-making. The further challenges in cultural heritage have to be coped with from an insight that lays out a useful response to changing technologies. This task has been carried out It is done via making decisions using game theory. This theory aids with decision-making and explains how particular outcomes can occur that are not optimal for the whole group. The collaborative investigation of research groups has been studied between a geography interdisciplinary research team and an art history research team in a GIS context. The stag hunt has been chosen as a tool to deal with this case.

This work has displayed that game theory via stag hunt yields a tool to address the collaboration between university departments with different scopes. It can derive a tool for strategic relations that take place in the training and application of GIS in cultural heritage. Game theory can operate as a base for cultural heritage professionals to adapt their scholarly methods to strategic cases and show modes to adjust the game parameters when the desired result is not the most expected. Future works will explore the role of game theory as a tool for cultural heritage specialists to accommodate their academic methods to handle new strategic circumstances from GIS technology challenges.

Acknowledgements This work was partially supported by Comunidad de Madrid (Spain) under the Multiannual Agreement with UC3M in the line of Excellence of University Professors (EPUC3M23), in the context of the 5th.\ Regional Programme of Research and Technological Innovation (PRICIT).

References

Arrighi C, Ballio F, Simonelli T (2023) A GIS-based flood damage index for cultural heritage. Int J Disaster Risk Reduction 90:103654. https://doi.org/10.1016/j.ijdrr.2023.103654

Aschenbach RJ (1991) Geographic information systems as a decision making tool. Ohio St LJ 52:351

Basir WNFWA, Setan H, Majid Z, Chong A (2014) Geospatial database for heritage building conservation. IOP Conf Ser Earth Environ Sci 18:012087. https://doi.org/10.1088/1755-1315/18/1/012087

Bonfanti I, Colucci E, De Ruvo V, Del Giudice M, Fasana S, Iacono E, Lingua AM, Matrone F, Ventura G, Zerbinatti M (2021) Development of an integrated BIM-GIS management tool for maintenance plan of historical heritage. In: Proceedings of the joint international event 9th ARQUEOLÓGICA 2.0 & 3rd GEORES, Valencia (Spain), 26–28 April. https://doi.org/10.4995/arqueologica9.2021.12131

Bruno N, Rechichi F, Achille C, Zerbi A, Roncella R, Fassi F (2020) Integration of historical GIS data in a HBIM system. Int Arch Photogram Remote Sens Spatial Inform Sci XLIII-B4:427–434. https://doi.org/10.5194/isprs-archives-xliii-b4-2020-427-2020

Burrough PA, McDonnell RA (1998) Principles of geographical information systems. Oxford University Press. http://dds.cepal.org/infancia/guide-to-estimating-child-poverty/bibliografia/capituloIV/Burrough%20Peter%20A%20y%20McDonnell%20Rachael%20A%20(1998)%20Principles%20of%20geographical%20information%20systems.pdf

Colucci E, Iacono E, Matrone F, Ventura GM (2022) A BIM-GIS integrated database to support planned maintenance activities of historical built heritage. Commun Computer Inform Sci 182–194. https://doi.org/10.1007/978-3-030-94426-1_14

Dhonju HK, Xiao W, Mills JP, Sarhosis V (2018) Share our cultural heritage (SOCH): worldwide 3D heritage reconstruction and Visualization via web and mobile GIS. ISPRS Int J Geo Inf 7(9):360. https://doi.org/10.3390/ijgi7090360

Espinola-Arredondo A, Muñoz-Garcia F (2023) Game theory. En Springer eBooks. https://doi.org/10.1007/978-3-031-37574-3

Huang Y (2024) Bibliometric analysis of GIS applications in heritage studies based on Web of Science from 1994 to 2023. Heritage Sci 12(1). https://doi.org/10.1186/s40494-024-01163-y

Liu B, Wu C, Xu W, Shen Y, Tang F (2024) Emerging trends in GIS application on cultural heritage conservation: a review. Heritage Sci 12(1). https://doi.org/10.1186/s40494-024-01265-7

Malinverni ES, Naticchia B, Garcia JLL, Gorreja A, Uriarte JL, Di Stefano F (2020) A semantic graph database for the interoperability of 3D GIS data. Appl Geomatics 14(S1):53–66. https://doi.org/10.1007/s12518-020-00334-3

Martínez-Hernández C, Piskorski R, Stoffelen A (2023) Designing online workshops for teacher trainees: heritage mapping with web GIS story maps. Eur J Geogr Eur J Geogr 14(3):68–78. https://doi.org/10.48088/ejg.c.mar.14.3.068.078

Matrone F, Colucci E, De Ruvo V, Lingua A, Spanò A (2019) HBIM in a semantic 3D GIS database. Int Arch Photogram Remote Sens Spatial Inform Sci XLII-2/W11:857–865. https://doi.org/10.5194/isprs-archives-xlii-2-w11-857-2019

Matrone F, Colucci E, Iacono E, Ventura GM (2023) The HBIM-GIS Main10ance platform to enhance the maintenance and conservation of historical built heritage. Sensors 23(19):8112. https://doi.org/10.3390/s23198112

Nishanbaev I, Champion E, McMeekin DA (2021) A web GIS-based integration of 3D digital models with linked open data for cultural heritage exploration. ISPRS Int J Geo Inf 10(10):684. https://doi.org/10.3390/ijgi10100684

Pedelì C (2013) An interdisciplinary conservation module for condition survey on cultural heritages with a 3D information system. Int Arch Photogram Remote Sens Spatial Inform Sci XL-5/W2:483–487. https://doi.org/10.5194/isprsarchives-xl-5-w2-483-2013

Pepe M, Costantino D, Alfio VS, Restuccia AG, Papalino NM (2021) Scan to BIM for the digital management and representation in 3D GIS environment of cultural heritage site. J Cult Herit 50:115–125. https://doi.org/10.1016/j.culher.2021.05.006

Rechichi F (2020) Chimera: a BIM+GIS system for cultural heritage. Int Arch Photogram Remote Sens Spatial Inform Sci XLIII-B4:493–500. https://doi.org/10.5194/isprs-archives-xliii-b4-2020-493-2020

Rickles, P., Ellul, C., & Haklay, M. (2017). A suggested framework and guidelines for learning GIS in interdisciplinary research. Geo, 4(2). https://doi.org/10.1002/geo2.46

Santos B, Gonçalves J, Martins AMT, Pérez-Cano T, Adell EM, Dimelli D, Lagarias A, Almeida PG (2021) GIS in architectural teaching and research: planning and heritage. Educ Sci 11(6):307. https://doi.org/10.3390/educsci11060307

Spiridon P, Ursu A, Sandu I (2016) Heritage management using GIS. In: 16th International multidisciplinary scientific GeoConference (SGEM 2016)

Tadelis S (2013) Game theory: an Introduction. Princeton University Press

Yagoub M, Yammahi AAA (2022) Spatial distribution of natural hazards and their proximity to heritage sites: case of the United Arab Emirates. Int J Disaster Risk Reduction 71:102827. https://doi.org/10.1016/j.ijdrr.2022.102827

Heritage Education and the Development of Digital Competencies in Pre-service Teachers Using the CatPat Collaborative App

Carlos D. Ciriza-Mendivil, Ana María Mendioroz Lacambra, and José Manuel Hernández de la Cruz

Abstract Information and communication technologies (ICTs), as learning tools, promote a shift towards new educational processes through didactic and pedagogical frameworks that go beyond traditional approaches. Their use transforms the concepts of interactivity and information empowerment. In this context, equipping future teachers with a wide range of tools, including apps, is essential for improving learning and raising awareness about cultural heritage. One such app, CatPat, was developed in 2022 by the Argos Research Group at the University of Zaragoza. This work examines its impact on pre-service primary school teachers at both the University of Zaragoza and the Public University of Navarra. The study uses an action-research methodology with a sample size of $n = 143$ and mixed-methods analysis employing custom-designed, validated questionnaires. The results align with the research objectives and demonstrate that students attribute significance to ICTs in heritage-related interaction and socialisation and recognise the CatPat app as a valuable tool for teaching and building heritage cybercommunities.

Keywords App · Digital competence · Heritage · Educational resource · Geolocation

---

C. D. Ciriza-Mendivil · A. M. Mendioroz Lacambra
Department of Human and Education Sciences, Grupo de investigación Aprendizaje a lo largo de la vida/Life-Long Learning Research Group, Public University of Navarra (UPNA), Navarra, Spain
e-mail: carlos.ciriza@unavarra.es

A. M. Mendioroz Lacambra
e-mail: anamaria.mendioroz@unavarra.es

J. M. H. de la Cruz (✉)
Department of Specific Didactics. ARGOS Research Group, IUCA. University of Zaragoza (UNIZAR), Zaragoza, Spain
e-mail: jmh@unizar.es

R. de Miguel González and P. Rivero Gracia (eds.), *Geospatial Technologies for Heritage Education*, Key Challenges in Geography,
https://doi.org/10.1007/978-3-031-96372-8_4

Introduction

Cultural heritage, as a social product, is continuously changing and evolving. Its value and meaning are attributed and shaped by the society and culture in which it exists. Therefore, the preservation and significance of these assets are deeply linked to the society and culture that interpret them (Muriel 2015; Fontal and Sánchez 2024). Heritage, as a social construct, is defined by the people who attribute value and meaning to it. It is a collective property that defines us as human beings sharing a common criterion, an established connection between individuals and communities regarding their cultural heritage (Fontal 2012). Heritage education, therefore, plays a pivotal role in knowing, socialising, and preserving cultural legacy.

Heritage, as a source of knowledge, gains relevance and meaning across different educational contexts and levels. Thus, heritage education is a form of curricular integration, helping students become bearers and transmitters of their legacy's values through the connections they form with heritage (Rodríguez 2018; Berríos et al. 2021). Cultural heritage acts as a bridge in the teaching and learning of history and geography, encompassing the development and use of basic technological competencies.

Information and communication technologies (ICTs), as learning tools, are key to overcoming traditional approaches to teaching and learning (Lengua et al. 2020). They favour changes in social perspectives, enhance communication through digital content, and empower individuals to develop their capacities, confidence, vision, and agency as a social group (Lévano et al. 2019).

The use of technology in education has taken on an increasingly central role, specifically in heritage education, where it has become a fundamental tool (Maldonado 2017; Jiménez and Cabero 2021). Effective teaching and appropriation of heritage involve understanding, appreciating, preserving, enjoying, transmitting, socialising, identifying, and raising awareness of it (Fontal and Sánchez 2024). Through these actions, technologies become instrumental in this process of creating heritage. In recent decades, the advantages and benefits of digital environments have elevated considerably the prominence and relevance of ICTs in heritage education (Simonelli 2019; Underberg and Hopp 2016). However, as Ciriza-Mendivil et al. (2022) highlight, teacher training in digital competences related to social sciences, including history, faces notable shortcomings. These authors argue that developing digital competence (DC) and awareness of cultural heritage is paramount for training future teachers.

In recent years, DC has drawn considerable attention, both in Europe and Spain. It is included as a key learning competence in European education policy (Commission of the European Communities 2006) and has been incorporated into the last three Spanish educational laws and into the curricula for initial teacher training in primary and early childhood education programmes (Ciriza-Mendivil et al. 2022). In today's hyperconnected world, teachers must effectively use technology to meet the demands of contemporary education (Fernández et al. 2018). Edge (2020) underscores the positive impact of technology on learning and its potential to bridge the digital divide.

DC acquires a broader meaning that is reflected in all dimensions inherent in teaching (García et al. 2022). For Pozos and Tejada (2018) and Herrera and Ochoa (2022), DC is a fundamental skill for individuals' social functioning. This prominence of DC has spurred extensive research on the topic from various perspectives and at different educational levels (Brenis et al. 2021; Padilla and Ayala 2021; Salazar and Lescano 2022; Reyna 2022; Viñoles et al. 2022; Montalvo et al. 2022; De la Cruz et al. 2023). While research on heritage learning in digital environments has grown, Fontal et al. (2024) note that most studies are exploratory, focusing on the implementation of innovation, which limits widespread adoption of the results.

The impact of geospatial technologies, geographic information systems, digital atlases, remote sensing, and geolocation-based mobile applications (apps) has increased in response to the challenges in education (Sebastián et al. 2023). Mobile apps in particular hold considerable potential as tools for teaching heritage (Vicent et al. 2020), as highlighted by several studies (Ibáñez and Kortabitarte 2016; Santacana et al. 2018; Kortabitarte et al. 2018; Luna et al. 2019). Based on all this, it is imperative to integrate these resources into the teaching practice from a socio-critical and symbolic approach, using active methodologies and introducing digital tools (Ciriza-Mendivil et al. 2023).

Ultimately, the integration of ICTs into new models of heritage education seeks to create a "scenario that supplements traditional face-to-face education, promoting the development of new ways of organising and accessing information and knowledge" (Ibáñez et al. 2018). Rubio's (2023) perspective aligns with this vision, emphasising that working with digital resources based on educommunication principles can encourage a dialogical, constructivist, and socio-critical approach to heritage in classrooms. Ibáñez et al. (2018) support the use of a wide variety of digital tools and resources, especially apps designed for heritage dissemination. Incorporating these tools into heritage teaching in teacher training programmes is essential, as they are key to knowing about, appreciating, and preserving cultural heritage (Gómez 2016). However, as noted by Ciriza-Mendivil et al. (2022), this can only be achieved if initial teacher training is prioritised as a strategic space for action, while acknowledging its inherent didactic and pedagogical complexities. In the university context, specifically in the training of future primary education teachers, the use of mobile technologies and digital resources for teaching social sciences and heritage education has yet to be fully established.

This research starts from the hypothesis that the CatPat app can contribute to pre-service teachers' knowledge and understanding of cultural heritage by supporting socialisation and interaction among users. It promotes the formation of heritage cybercommunities and advocates the preservation of cultural legacy by stimulating the community's active participation in valuing and preserving it.

To this end, the general objective (GO) of this study is to analyse the impact of the CatPat app, developed in 2022 by the Argos Research Group at the University of Zaragoza (UNIZAR), on pre-service primary school teachers at both UNIZAR and the Public University of Navarra (UPNA). This work focuses on three specific objectives: Specific Objective 1 (SO1), to develop a didactic proposal in classrooms based on the open-access and collaborative CatPat app; Specific Objective 2 (SO2), to analyse the impact of the didactic proposal based on CatPat on the appreciation of heritage; and Specific Objective 3 (SO3), to assess the app's influence in fostering interest in heritage-related digital applications.

Methodology

The study used the action-research methodology. As Latorre (2007) explains, this method aims to transform and understand educational practice by approaching reality and connecting knowledge and change. The sample consisted of $n = 143$ students of the primary education teaching degree programmes at UNIZAR (third year) and UPNA (second year), selected by non-probabilistic intentional sampling. All students consented to participate. Data collection involved both quantitative and qualitative analyses, using a pre- and post-test format based on an ad hoc questionnaire validated by the Ethics, Animal Experimentation and Biosecurity Committee of the UPNA. The design of this questionnaire was informed by the work of Fernández et al. (2021) and the analysis by Ciriza-Mendivil et al. (2022) regarding the assessment and development of heritage education competences in university students. Key insights into Spanish students' perceptions of heritage came from the work of Marín and Fontal (2020), while Fontal et al. (2020) provided valuable findings on the social significance of heritage.

The study employed a pre-test/post-test design, administering two separate questionnaires before and after the intervention, respectively. The first explored the relevance of cultural heritage, and the second evaluated specifically the CatPat app.

The pre-test sought to understand students' perceptions of cultural heritage, heritage education, and the use of apps for teaching, socialising, and preserving cultural legacy. It consisted of 13 mandatory items, including both open- and closed-ended Likert-type questions with five response options. A total of $n = 70$ students completed the questionnaire.

The post-test consisted of 24 items, including both open- and closed-ended Likert-type questions. A total of $n = 73$ students completed the questionnaire, which was administered after the pre-service teachers had used the CatPat app. The post-test aimed to assess students' perceptions of the significance of cultural heritage, comparing them to pre-test data. It also sought to determine the value and usefulness they attributed to the app for teaching, socialising and preserving cultural legacy, and forming heritage cybercommunities.

Results

We present the results in two sections based on the study's nature and objectives. First, we analyse the items included in both the pre-test and post-test to assess changes in pre-service teachers' perceptions of the significance and value of heritage and heritage education before and after using the CatPat app. Second, we examine the items related specifically to the use of CatPat in this study, including pre-service teachers' evaluations, perceptions, and uses of the tool.

Item 1 of the questionnaire asked an open-ended question: "What do you understand by cultural heritage?" Pre-test responses were highly varied, with 60% associating cultural heritage with artistic value and 20% linking it to the legacy and inheritance of past generations. Fewer responses mentioned customs, traditions and, in two cases, natural heritage. Post-test responses, after the introduction of the CatPat app, showed greater uniformity and complexity. For instance, 41% defined cultural heritage as valuable to a community, 20.3% described it as something with artistic and historical value inherited from the past, and 38.7% saw it as encompassing everything that represents us, has historical and artistic value, and must be preserved.

Item 2 asked pre-service teachers about their knowledge of heritage, and item 3 requested them to provide six pertinent examples. In both cases, responses from both the pre-test and post-test mostly linked heritage to material assets, especially monuments (50%). While 34.4% of respondents recognised both tangible and intangible assets and 15.6% included natural components, the examples provided in item 3, in both the pre-test and post-test, consistently focused on built material cultural heritage, with 87.3% of students identifying monuments, Spanish cities, and iconic works of cultural heritage from other European and American countries. Only 8.5% mentioned natural heritage (e.g., the Pyrenees and Ordesa Valley) and 4.2% cited intangible heritage (e.g., dry stone technique, the drum festivals of Aragon, the Way of Saint James, and the San Fermín festivities in Pamplona) (see Table 1).

Table 1 Index of answers to questions 1, 2, and 3 of the analysis instrument (post-test)

Question 1		Question 2		Question 3	
What do you understand by cultural heritage?		What type of wealth do you know?		Six relevant examples of heritage	
%	Answers	%	Answers	%	Answers
41	That which has value	50	Heritage Immovable material	87.3	tangible cultural heritage in which monuments and cites are identified
20.3	What has artistic and historical value and we inherit from the past	34.4	Heritage made up of tangible and intangible assets	8.5	Exponents of natural heritage
38.7	Everything that represents us, has historical and artistic value and identifies a specific culture and must be preserved	15.6	Recognize the natural component as part of the heritage	4.2	Exponents of immaterial heritage

Source Own elaboration

Item 4 analysed pre-service teachers' perceptions of the relevance of cultural heritage (Fig. 1). This item showed one of the most outstanding improvements following the use of the CatPat app. In the pre-test, 55% of the students considered cultural heritage "very important", 40% "important", and 4.3% did not have a clear opinion.

After implementing the CatPat app, pre-service teachers' perceptions of the importance of heritage changed markedly. Nearly all respondents (94%) viewed heritage as "very important" and only 5% as simply "important" (Fig. 2). This represents a considerable improvement from the pre-test results. The percentage of responses considering heritage as "very important" increased significantly. However, the low proportion of pre-service teachers unsure of how to respond in the pre-test disappeared entirely.

These results confirm the effectiveness of the CatPat app in promoting cultural heritage's value as an "important" or "very important" educational resource. With 94.5% of the sample now considering cultural heritage "very important" and 5.5% "important", for 100% of the pre-service teachers who developed a didactic proposal using the CatPat app heritage gained outstanding relevance within the educational sphere.

Both questionnaires show that pre-service teachers place substantial value on heritage education. In the first questionnaire, 51.4% of respondents considered heritage education "very important", while 42.9% deemed it "important". Combining these two categories, we can conclude that 94.3% regarded heritage education as relevant. The results are even more encouraging in the post-test responses, as 95.5% of respondents rated heritage education as "very important" and only 4% considered it "important" (Figs. 3 and 4).

The second section of the questionnaire, only included in the post-test, evaluated the CatPat app and revealed that 98.7% of students considered the app interactive. Only 1.4% were not sure about the app's level of interactivity. Overall, the

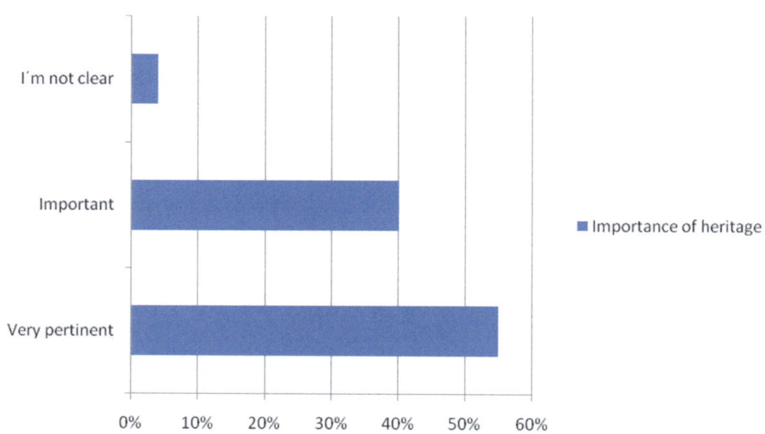

Fig. 1 Pre-test importance of heritage. *Source* Own elaboration

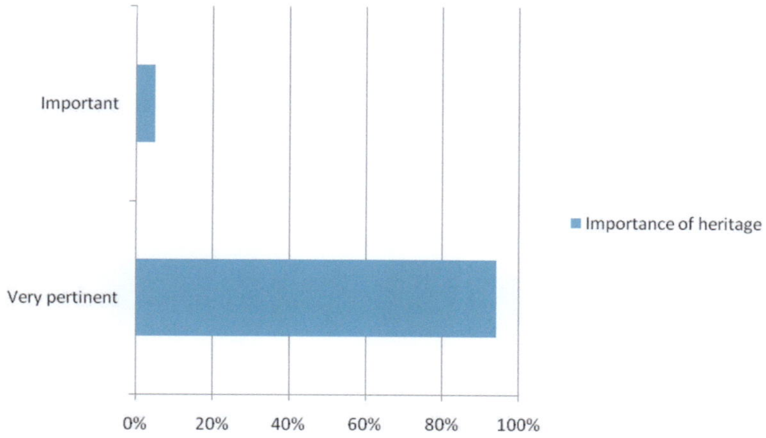

Fig. 2 Post-test importance of heritage. *Source* Own elaboration

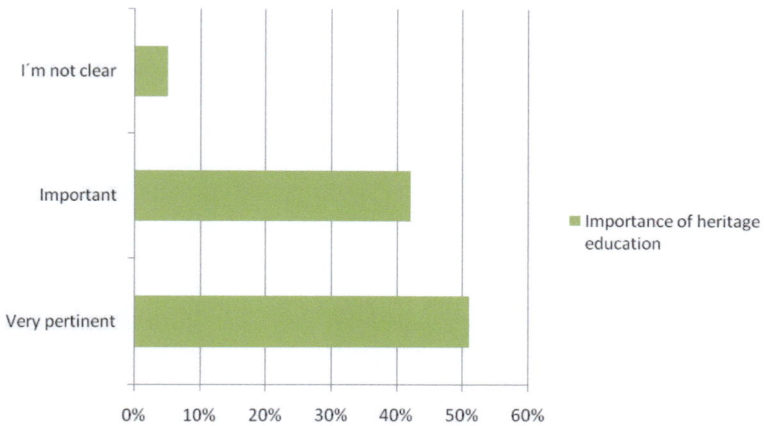

Fig. 3 Pre-test importance of heritage education. *Source* Own elaboration

responses indicated a positive inclination towards using the app and its potential for implementation in teaching.

As for the possibilities of CatPat, 98.6% of respondents agreed that the app is effective for teaching, socialising, valuing, and preserving cultural heritage. Notably, 100% of the sample concurred that it is applicable to teaching, nurtures respect, helps users discover or learn about new heritage sites, and fosters an interdisciplinary approach to heritage. Additionally, 97.3% of respondents believed CatPat develops a sense of belonging and identity. Thus, the entire sample acknowledged that the app was an important tool for raising awareness, sharing content, cultivating dialogue, and promoting awareness and knowledge about heritage (Fig. 5).

Nearly all respondents (98.7%) found CatPat useful for working with social sciences curriculum content. The entire sample acknowledged that CatPat facilitates

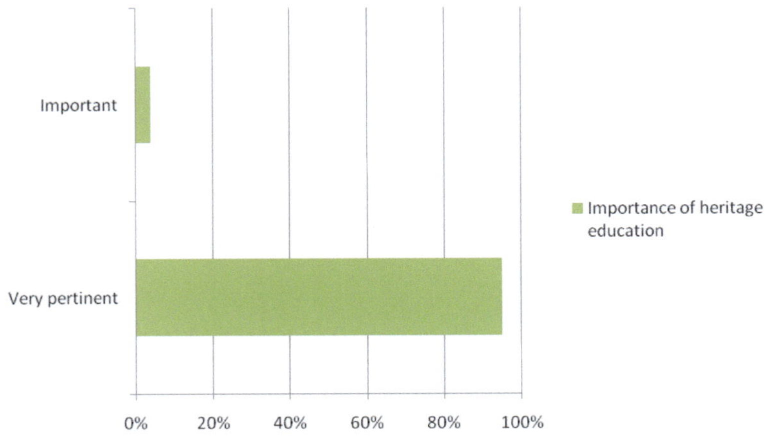

Fig. 4 Post-test importance of heritage education. *Source* Own elaboration

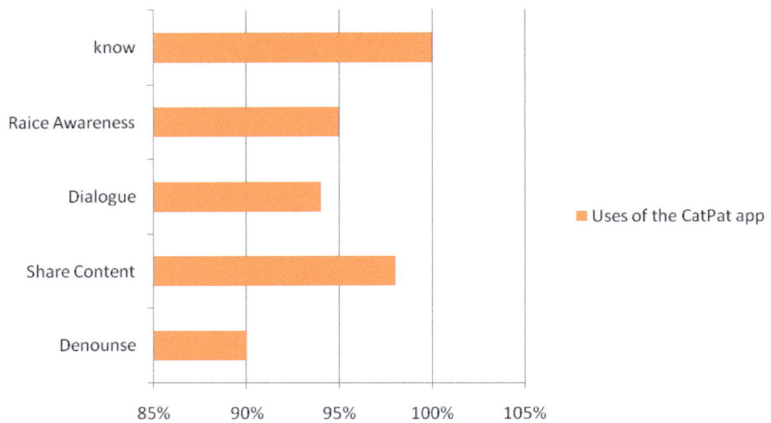

Fig. 5 Post-test about using the CatPat app. *Source* Own elaboration

the development of some of the specific competences (SC) included in the current primary education curriculum.

While the pre-test responses revealed that the student teachers believed in developing many of the SC in the social sciences curriculum through the use of apps (Fig. 6), the post-test responses demonstrated greater specificity (Fig. 7). In other words, instead of an increase in the number of competences identified, the responses focused on fewer competences, more closely linked to heritage education. This shift indicates that, as students became more informed and aware, they associated heritage education using CatPat with SC such as SC 1, 2, 5, 6, 7 and 8, primarily related to using digital resources and tools safely and responsibly; solving simple scientific problems; identifying characteristics of natural, social and cultural environments;

recognising the causes and effects of human intervention in the environment; understanding the traits of different societies over time and the roles that human beings have played in them; recognising and valuing diversity and gender equality; and respecting other cultures.

Finally, 57.9% of the sample showed a positive perception of using social media, ICTs, and heritage education resources in the classroom. However, 40% were either unaware or unsure about the possibility of connecting social media, digital competences, and heritage (Fig. 8). This response revealed a lack of knowledge about the

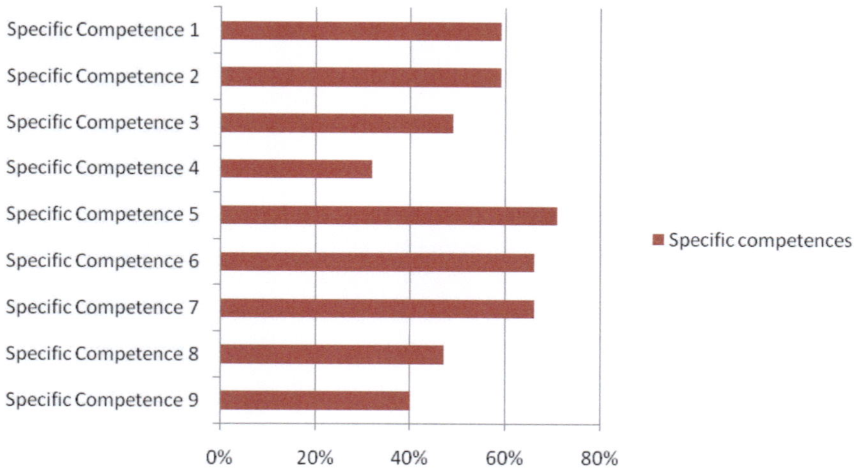

Fig. 6 Pre-test development of SC. *Source* Own elaboration

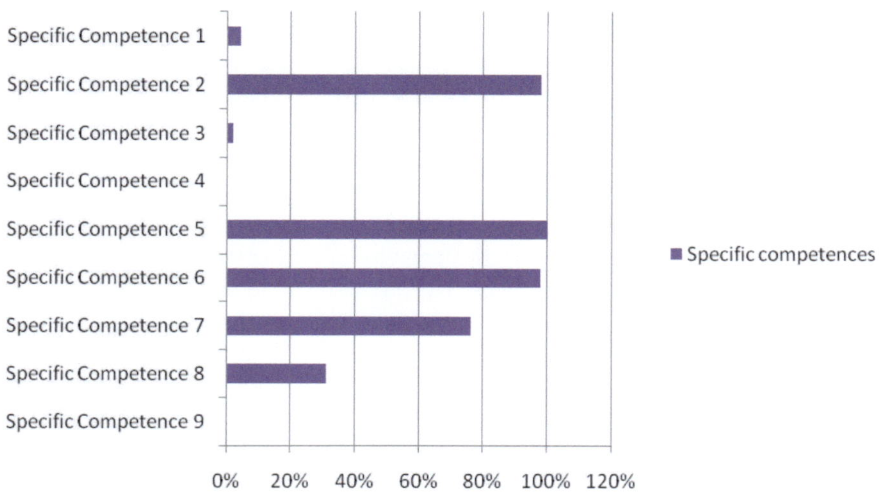

Fig. 7 Post-test development of SC. *Source* Own elaboration

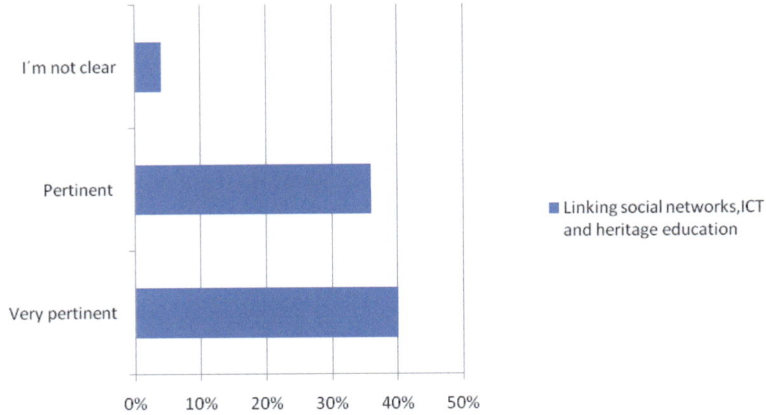

Fig. 8 Relationship between social media, ICTs, and heritage education. *Source* Own elaboration

applicability of social media and ICTs in heritage education for developing digital competences and raising students' awareness of cultural heritage. These findings underscore the need for specific proposals, like the one presented in this study, to leverage the potential of resources such as CatPat in teaching practice through interdisciplinary and multidisciplinary strategies.

Discussion

Cultural heritage, a source of knowledge enriched by the transmission of social, cultural, and economic values, must be preserved as a vital social resource (Fontal 2012). Training future teachers in its significance and the methods and tools for understanding and appreciating it is crucial. Heritage resources in particular are highly suitable for teaching and learning history and geography, while also cultivating basic technological skills. This aligns with the growing challenges and advancements in social media and ICTs. Developing digital competences through the use of technological resources remains a challenge in teaching and learning across various educational levels and contexts (Ciriza-Mendivil et al. 2022).

This study aimed to address this challenge by using the CatPat app in primary education teaching degree programmes. The proposal and its implementation in the classrooms contributed to the development of skills in the use of resources for heritage education (OE1). Students observed the value of heritage in teaching, socialising, appreciating, raising awareness, and understanding. Additionally, as a geolocation tool for cultural and natural heritage, CatPat favoured geographical education and the development of spatial citizenship (Sebastián et al. 2023) (Fig. 9).

Fig. 9 Images by CatPat. *Source* CatPat screenshots

The results align with the perspectives offered by Ciriza-Mendivil et al. (2022), Ibáñez et al. (2018), and Maldonado (2017), who supported the integration of technologies as new models for heritage education and a means of teaching and appreciating cultural heritage by creating supplementary educational spaces. The implementation of this approach in teaching showed students that the use of technology provides access to and offers opportunities for informal learning. The CatPat app has proven to be a valuable tool for heritage learning in diverse educational contexts, promoting values, creating alternative learning environments, fostering heritage-focused cybercommunities, and motivating student engagement (Herrera and Ochoa 2022).

The pre-test and post-test results related to OE2 demonstrated students' recognition of heritage's importance, aligning with Fontal and Sánchez (2024) and Muriel (2015). While the study highlighted differences between the pre-test and post-test, revealing an initial bias with a negative inclination towards heritage and its value, most respondents ultimately acknowledged and appreciated heritage education. Based on the responses, students understood the crucial role of heritage education across various levels and contexts, seeing it as a means to help pupils become bearers and transmitters of cultural values through the connections they form with their cultural legacy. This perspective supports the findings of Rodríguez (2018) and Berríos et al. (2021).

The study also emphasised a lack of resources for teaching heritage education and developing digital competences, an issue that this study sought to address. Regarding CatPat, this work revealed a lack of familiarity with such resources, but also a growing interest in digital heritage apps. Over half the sample acknowledged the high applicability of these digital tools in teaching, particularly CatPat (Sebastián et al. 2023).

Pre-service teachers viewed CatPat as an attractive and motivating tool (Luna et al. 2019) that can and should be integrated into educational practice (Gómez 2016). For students, the app served as an interactive medium, which is consistent with García and Rubio's (2023) assertion that CatPat develops social interactions and connections and is a repository of heritage.

Although the results indicate a positive attitude towards using social media and ICTs in heritage education, the pre-test revealed some confusion about these tools. The responses also highlighted a lack of knowledge among students, which contributed to their initial reluctance to adopt certain resources. However, post-test data showed a clear positive shift. Familiarity with CatPat and its usefulness prompted pre-service teachers from UNIZAR and UPNA to adopt a more favourable view of ICTs and social media in heritage education. Furthermore, the study demonstrated how the app enhanced participants' digital teaching competencies and improved their understanding of its implications for developing specific curricular aspects.

Conclusions

This study demonstrates that pre-service teachers of the primary education teaching degree programme perceive heritage and heritage education as valuable sources of knowledge, essential for shaping engaged citizens who critically reflect on their social reality and environment. However, it also highlights that the lack of specific proposals and knowledge of digital resources often hinders the use of digital tools in heritage education.

The CatPat app promoted new spaces for building knowledge related to DC and heritage education. The results are encouraging regarding the app's effectiveness in developing DC and achieving a more complex and profound teaching competence. The data collected indicate the app's ease of use and potential for discovering new heritage, creating dialogues, geolocating cultural and natural resources, and forming heritage-based cybercommunities that contribute to the socialisation and understanding of the environment and its components.

The study also showed that CatPat plays a key role in developing digital skills and raising awareness about cultural heritage by facilitating fundamental heritage-related actions—knowing, understanding, valuing, raising awareness, preserving, enjoying, transmitting, socialising, and identifying heritage. In addition, the app promotes spatial thinking, spatial citizenship and geographic, historical, and social identity awareness.

Finally, it is crucial for future teachers to receive heritage training that fosters connections with heritage from a symbolic-identity perspective, ensuring knowledge about and preservation of heritage while leveraging the potential of ICT resources. While the small sample size limits the conclusions, this study opens the door to further research in classrooms involving students from other universities. This would enable us to assess CatPat's impact across various educational contexts and to explore possibilities to create heritage cybercommunities that cultivate awareness and communication around cultural heritage.

Funding This research was also possible thanks to the project "Participatory Digital Tools for Supporting Heritage Cybercommunities" TED2021-131174B-I00, financed by MICIU/AEI/ 10.13039/501100011033 and by FEDER A Way to Make Europe, and to the ARGOS research group of the University of Zaragoza, S50_23R, Government of Aragon 2023–2025. The research outcome is derived from a research stay of José Manuel Hernández de la Cruz at the Public University of Navarra (UPNA).

References

Berríos A, Tessada V, Gallegos F (2021) Proposals for an heritage education model in initial teacher training of pedagogy in history, geography, and social sciences. Revista Brasileira de Educação 26:e260029. https://doi.org/10.1590/S1413-24782021260029

Brenis A, Alcas N, Maldonado F (2021) El desarrollo de competencias digitales en docentes universitarios frente al auge de la educación virtual. 593 Digital Publisher CEIT 6(4):111–121. https://doi.org/10.33386/593dp.2021.4.651

Ciriza CD, Lacambra AM, Hernández JM (2022) Technological pedagogical content knowledge: implementation of a didactic proposal for preservice history teachers. Front Educ 7:852801. https://doi.org/10.3389/feduc.2022.852801

Ciriza-Mendívil CD, Mendioroz A, Hernández de la Cruz JM, Rivero P (2023) El Patrimonio Cultural en la Educación Infantil. Análisis de la Percepción del Profesorado en Formación. Social Educ Hist 12(3):274–291. https://doi.org/10.17583/hse.12025

Comisión de las Comunidades Europeas (2006) Recomendación del Parlamento Europeo y del Consejo (18 de diciembre) 2006/962/CE. Comisión Europea, Bruselas

De La Cruz J C, Santos MJ, Alcalá MJ, Victoria JJ (2023) Competencias digitales docentes en la educación superior. Un análisis bibliométrico. Hachetetepé. Revista científica De Educación Y Comunicación 26:1103. https://doi.org/10.25267/Hachetetepe.2023.i26.1103

Foundation E (2020) The impact of Covid-19 oneducation: evidence on the early impacts oflockdown. Edge Foundation, London

Fernández AR, Fernández M & Medina ID (2021) The evaluation of the competences in Heritage Education of students of the Degree of History of the University of Malaga. IJNE Int. J. New Educ. , 6, 1–17. https://doi.org/10.24310/IJNE3.2.2020.8451

Fernández E, Leiva JJ, López E (2018) Competencias digitales en docentes de Educación Superior. Revista Digital de Investigación en Docencia Universitaria 12(1):213–231

Fontal O (2012) Patrimonio y educación: una relación por consolidar. Aula de Innovación Educativa, núm. 208:10–13. https://hdl.handle.net/11162/87484

Fontal O, Sánchez I (2024). Personas y patrimonios compartidos Conceptualización del patrimonio a partir de relatos en entornos digitales Perfiles Educativos XLVI:183. https://doi.org/10.22201/iisue.24486167e.2024.183.59070

Fontal O, Ibañez-Etxeberria A, Arias VB, Arias B (2024) Q-Herilearn: assessing heritage learning in digital environments. A mixed approach with factor and IRT models. PLOS One 19(3). https://doi.org/10.1371/journal.pone.0299733

Fontal O, Martínez M, Cepeda J (2020). Social signification of heritage: An analysis of heritage perception in the Community of Madrid. Aula Abierta 49:17–24. https://doi.org/10.17811/rifie.49.1.2020.17-24

García M, Morales, MJ, Gisbert M (2022) El desarrollo de la Competencia Digital Docente en Educación Superior. Una revisión sistemática de la literatura. RiiTE Revista interuniversitaria de investigación en Tecnología Educativa, 173–199. https://doi.org/10.6018/riite.543011

García-Ceballos S, Rubio A (2023) ¿Cuántos somos necesarios? CATPAT, aplicación colaborativa para la creación de cibercomunidades patrimoniales en contextos escolares. Revista PH n.º 109:166–168. https://dialnet.unirioja.es/servlet/articulo?codigo=8997508

Gómez I (2016) La inclusión de las tecnologías en la formación inicial del profesorado: una intervención de aula a través del modelo TPACK. Tendencia. Pedagogo. 28:133–152. https://doi.org/10.15366/tp2016.28.010

Herrera JC, Ochoa ED (2022) Análisis de la relación entre educación y tecnología. Cultura Educación Sociedad 13(2):47–68. https://doi.org/10.17981/cultedusoc.13.2.2022.03
Ibáñez A, Kortabitarte A (2016) Apps, Redes Sociales y dispositivos móviles en educación patrimonial. EUSKOIKER INFORME FINAL_02dec_v5 (cultura.gob.es)
Ibáñez A, Fontal O, Rivero P (2018) Educación patrimonial y TIC en España: marco normativo, variables estructurantes y programas referentes. Arbor 194(788):a448. https://doi.org/10.3989/arbor.2018.788n2008
Jiménez MJ, Cabero J (2021) Los conocimientos tecnológicos, pedagógicos y de contenidos del profesorado universitario andaluz sobre las TIC. Análisis desde el modelo TPACK. Innoeduca Int J Tecnología Educativo. Innovación 7:4–18. https://doi.org/10.24310/innoeduca.2021.v7i1.11940
Kortabitarte A, Gillate I, Luna U, Ibáñez A (2018) Las aplicaciones móviles como recursos de apoyo en el aula de Ciencias Sociales: estudio exploratorio con el app" Architecture gothique / romane" en Educación Secundaria. Ensayos, Revista de la Facultad de Educación de Albacete 33(1):65–79
Latorre A (2007) La investigación- acción. Conocer y cambiar la práctica educativa. Grao, Barcelona, España
Levano L, Sanchez S, Guillén P, Tello S, Herrera N, Collantes Z (2019) Competencias digitales y educación. Propósitos Y Representaciones 7(2):569–588. https://doi.org/10.20511/pyr2019.v7n2.329
Lengua C, Bernal G, Flórez W, Velandia M (2020) Tecnologías emergentes en el proceso de enseñanza aprendizaje: hacia el desarrollo del pensamiento crítico. Revista Electrónica Interuniversitaria de Formación Del Profesorado. https://doi.org/10.6018/reifop.435611
Luna U, Ibáñez A, Rivero P (2019) El patrimonio aumentado. 8 apps de Realidad Aumentada para la enseñanza-aprendizaje del patrimonio. Revista Interuniversitaria De Formación Del Profesorado. Continuación De La Antigua Revista De Escuelas Normales
Maldonado S (2017) Educación patrimonial y redes sociales. De la investigación a la acción: El proyecto Educación Patrimonial. Pulso. Revista De educación, 40:97–111. https://doi.org/10.58265/pulso.5116
Marín S, Fontal O (2020) Secondary Education student's perceptions around Heritage. Arte Individuo Soc. 32:917–933. https://doi.org/10.5209/aris.64657ARTÍCULOS
Montalvo VM, Villena MP, Franco GK (2022) Competencias digitales en docentes del Perú. Alpha Centauri 3(2):14–21. https://doi.org/10.47422/ac.v3i2.75
Muriel D (2015) La mediación experta en la construcción del patrimonio cultural como producción contemporánea de 'lo nuestro. AIBR. Revista de Antropología Iberoamerica-na 10(2):259–288
Padilla JC, Ayala GG (2021) Competencias digitales en profesores de educación superior de Iberoamérica: una revisión sistemática. RIDE. Revista Iberoamericana para la Investigación y el Desarrollo Educativo 12(23):e056.https://doi.org/10.23913/ride.v12i23.1096
Pozos KV, Tejada J (2018) Competencias Digitales en Docentes de Educación Superior: Niveles de Dominio y Necesidades Formativas. Revista Digital de Investigación en Docencia Universitaria **12**(2):59–87. https://doi.org/10.19083/ridu.2018.712
Reyna A (2022) Competencias digitales y desempeño docente en los colegios de Latinoamérica. Desafios 13(1):25–36. https://doi.org/10.37711/desafios.2022.13.1.367
Rodríguez ME (2018) La Educación Patrimonial y la Formación Docente desde la Transcomplejidad. TELOS. Revista de Estudios Interdisciplinarios en Ciencias Sociales **20**(3):431–449. https://dialnet.unirioja.es/descarga/articulo/6553270.pdf
Rubio A (2023) CATPAT, una app de mapeado colaborativo: Análisis de aplicaciones digitales para la educomunicación patrimonial en contextos educativos formales. [Trabajo final de asignatura, Universidad de Zaragoza]. En prensa
Santacana J, Asensio M., Brouard, López V, Martínez T (Coords.) (2018) La evaluación de las apps en el patrimonio cultural": Somonte-Cenero, Gijón (Asturias): Trea, [ISBN: 978–8417140397]

Salazar MR, Lescano GS (2022) Competencias digitales en docentes universitarios de América Latina: Una revisión sistemática Alpha Centauri 3(2):02–13. https://doi.org/10.47422/ac.v3i2.69
Sebastián M, Kratochvíl O, De Miguel R (2023) Educación Geográfica y Ciudadanía Espacial: Cartografía colaborativa para el aprendizaje del entorno local en un contexto global. In: Klonari A, Lázaro D, Torres ML, Kizos A (eds) Re-visioning Geography. Principales desafíos de la geografía. Springer, Cham. https://doi.org/10.1007/978-3-031-40747-5_10
Simonelli MR (2019) Integración de las tic en las Ciencias Naturales para el desarrollo de las competencias: caso UPEL-IPMAR. Investigación y Postgrado 34(2):9–37. https://dialnet.unirioja.es/descarga/articulo/7979548
Underberg N, Hopp C (2016) Investigación del aprendizaje cultural en entornos digitales a través de Perudigital. Etnografía Vis 5(2):19–34. https://doi.org/10.12835/ve2016.2-0065
Viñoles V, Sánchez A, Esteve FM (2022) Desarrollo de la competencia digital docente en contextos universitarios. Una revisión sistemática. REICE. Revista Iberoamericana sobre Calidad, Eficacia y Cambio en Educación 20(2):11–27. https://doi.org/10.15366/reice2022.20.2.001
Vicent N, Gillate I, Campos T, Castrillo J (2020) Potencial educativo de apps de Memoria Histórica. Estudio de casos. Aula abierta 49(1):35–43. https://doi.org/10.17811/rifie.49.1.2020.35-44

Coastal Landscapes with Heritage Features to Boost Spatial Thinking Through a Place-Based Learning Workshop

Alfonso Garcia de la Vega, Marcos Chica Diaz, and Keumbee Lee

Abstract The chapter shows the design, approach and results of an educational research project on the acquisition of geographic and spatial concepts together with heritage concepts for K-12. The project topic was the sixteenth-century watchtowers on the Mediterranean coast. A workshop and a fieldtrip were carried out. In the workshop, searches were done for information related to the location and construction of the watchtowers through online cartographical resources and reading of historical, literary and cartographical documentation. Fieldtrips were then made for observation and relationship among the watchtowers and the coastal landscape. From these activities, spatial problems were raised. Both problems are related to location, orientation and position of natural and cultural features and problems with the spatial interrelations between the watchtowers and the coastal landscape. The results offer some shortcomings in the training of students linked to the use of online maps, reading of cartographical documentation and the ability to represent simple cartographic schemes. However, students reinforced these shortcomings during the workshop and schools. The excitement with the applied use of online maps and the experience of the fieldtrip reinforced the acquisition of spatial concepts in relation to heritage. In this sense, heritage elements play an added value to the acquisition of geographical content and spatial geoskills.

Keywords Place-based learning · Spatial thinking · Geographical workshop · Mazarrón Bay

Introduction

Every day, we all use spatial skills to perform routine tasks. When traveling by public transport or by bicycle, you may need to decide which bus or metro station is closest to your workplace. Navigation, especially with mobile devices, is arguably

A. Garcia de la Vega (✉) · M. Chica Diaz · K. Lee
Universidad Autónoma de Madrid, Madrid, Spain
e-mail: alfonso.delavega@uam.es

R. de Miguel González and P. Rivero Gracia (eds.), *Geospatial Technologies for Heritage Education*, Key Challenges in Geography,
https://doi.org/10.1007/978-3-031-96372-8_5

more intuitive than reading paper maps. In fact, scale and distance are constantly changing, regardless of whether you are making a daily commute or taking a trip across the country. Even mobile devices incorporate a variable factor that maps lack time. Apps offer tracks conditioned by the quickest and shortest route. Otherwise, you must specify the milestones you want to visit, regardless of the time that this takes.

The project was acquiring knowledge on historical facts and spatial concepts and setting geographical relationship between landscape and heritage features. During the fifteenth and sixteenth centuries, the Spanish Mediterranean coast, especially alongshore Valencia and Málaga, endured numerous Berber incursions from North Africa. The Berbers, who had become pirates and corsairs, carried out numerous raids both at sea and on the coast. Farmers, miners and fishermen were left defenceless on the Mediterranean coast during these centuries. This fact led to a geographical reconnaissance of the coast to establish coastal watchtowers on the coast of Murcia. The project scope is limited to the coastal section between Cabo de Palos (Cartagena) and Cabo Cope (Águilas) (Fig. 1).

With the help of cartographic viewers (Iberpix) and historical and current online topographic cartography (GoogleMaps), possible hypotheses were proposed regarding the economic attractiveness of the coast around the sixteenth century. Next, fieldtrips around the watchtowers were made, so that the relationship between the location and the defended maritime and coastal space could be observed and verified. To do this, it was intended to pose spatial problems on the surface of the defended space, the spatial interrelations of the watchtowers to cover the maximum defended space. Likewise, these observations allowed us to recognize linear distances to the coast and altitudes in the relief forms of the coast (Figs. 1 and 2).

However, the growth, expansion and power of Geospatial Technology require citizens with well-developed spatial thinking skills. This chapter aims to describe the objectives of a project whose underlying framework was spatial and analogical thinking in education. To do so, firstly, the origin of this project is explained. The project is a consequence of trying to link scientific and school knowledge. The research group has been participating in activities to disseminate fieldtrips and workshops. In these, paper and online cartography have always been used, as well as

Fig. 1 City map of Spanish Mediterranean coast, Mazarrón Bay and Inlet and Tiñoso Cape (Cartagena, Murcia). *Source* Instituto Geográfico Nacional (Spain), 2024

Fig. 2 Panoramic view La Azohía. At the end of the coast is the Santa Elena Watchtower, Tiñoso Cape (Cartagena, Murcia). *Source* A. Garcia de la Vega, 2024

literary and artistic documents. From all this arises the interest in exploring spatial thinking expressed in a narrative workshop, where one can draw, map and express with clay what is visualized and experienced.

Therefore, the method and the two educational techniques of the project are described below. The method consists of the design of educational research. The two key processes of this project are the narrative workshop and the fieldtrip. In the workshop, the ability to represent known space (coast, beach, cliff, hill, bay, inlet and gulf) has been arranged. During the fieldtrip, the ability to identify relief features and heritage landmarks, as well as to establish analogies between them, was developed. The results are then presented and discussed, highlighting the findings of the exploration of spatial and analogical thinking. Finally, the chapter concludes with the possibilities of replicating a project of this type by promoting GIS in better social conditions. And, above all, highlighting the role that heritage could offer both as a narrator of a curricular topic and as a supporter of spatial geoskills.

Background

An important aspect of many scientific discoveries stems from the spatial nature of the relevant data, states Newcombe (2016). This author, and also Gersmehl (2014), reinforces this statement with the well-known case of cholera in London in the mid-nineteenth century. Both Gersmehl (2014) and Newcombe (2016), from different disciplinary perspectives, highlight the role of spatial thinking in science. Thus, it is worth highlighting the findings of doctor John Snow in the Soho neighbourhood in

London. On the map, Snow marked the water pump that caused cholera in the neighbours. For all this, Bednarz et al. (2022) affirm that there is an emerging consensus that spatial thinking is fundamental to later success in mathematics, science, and geography. Thus, it is also considered that didactic strategies to enhance spatial thinking instruction, including using a spatial thinking taxonomy, teaching resources, and educational activities to assess geospatial skills, are discussed. After showing the results of numerous empirical tests, Uttal et al. (2013) point out that space education promotes STEM.

In the project, both spatial thinking and analogical thinking come together for the development of the geographic narrative workshop and the fieldtrip. Spatial thinking is promoted by reading online cartography through a cartographic viewer. In addition, giving emotional meaning and heritage value to the watchtowers, close to their homes, by seeking information on internet search engines. And, finally, the spatial experience during the fieldtrip, which involved promoting spatial navigation skills, such as distance and orientation. These steps correspond to those marked by Newcombe (2016).

The watchtowers become the didactic framework to understand the curricular contents of the environment and to unravel the skills of spatial thinking. Now, it is about enhancing spatial skills with the technological means available. And, above all, linking the watchtowers in a real spatial context (Mazarrón Gulf) and, consequently, exploring the development of simple and complex geoskills. On the one hand, spatial thinking has numerous specific studies on the identification of spatial skills, from a geographical perspective (Gersmehl 1991, 2014; Metoyer et al. 2015; Sinton et al. 2013; Sinton 2016). All these authors point to the elements of cartographic language, which require an educational and didactic boost for spatial thinking skills. For his part, Gersmehl (1991) recognizes certain cartographic concepts, such as location, distance, direction, area and volume, scale and representation. These concepts could be classified into two types those dedicated to cartography (location, scale and representation) and others dedicated to navigation (distance and direction), according to Sinton (2016).

On the other hand, in line with the assertion by Metoyer et al. (2015) that spatial thinking has always been a fundamental cognitive skill for competency in geography, the special capacity in the context of geospatial technologies (GST) needs to be further developed. Furthermore, to educate a citizenry committed to the use of technologies, as promoted by Jekel et al. (2015) through the term Spatial Citizenship. This commitment points towards the use and application of spatial thinking and its educational resources towards the resolution of everyday and social problems. That is, from navigating with a mobile device to getting to a place of solving problems related to uncontrolled urbanism. To be precise, Goodchild and Janelle (2010) state that recognition that a spatial perspective can lead to valuable insights and the development of new theory about social and environmental phenomena distributed over the surface and near-surface of the Earth.

Cognitive processes associated with spatial thinking demand wide-ranging mental abilities. The simple skill set particular to the geospatial thinking domain is to observe, identify, relate and compare. Instruction momentum stirs complex thinking abilities that dictate reasoning strategies linked to classification, hierarchy and analogy for problem framing and solving at different scales and by varied geospatial resources (Garcia de la Vega 2019). Analogical thinking triggers relationships of similarity applied to environmental spatial thinking. Besides, this sort of reasoning benefits from prior experiences that enable the citizen to link specific, varying and comparable situations or contexts (Holyoak 1995; Holyoak and Thagard 1997). Analogical thinking activates similarity relationships applied both to knowledge related to the coastal watchtowers and to the spatial environment. In fact, the students know the environment well, but they are unaware of its spatial, geographical and historical features. This means that the students were able to visit them with their families. However, the students are unaware of the causes and time of their construction, the distance between them, the spatial references with the coast and the bay, for example.

To all this, it should be added the educational value of geospatial technologies, due to their application as a teaching resource. Thus, Wilson (1996) believes that learning with technologies requires collaborative interaction embedded in a technology-enhanced constructivist learning environment. He underlines that "technology-enhanced constructivist learning currently focuses on how representations and applications can mediate interactions among learners and natural or social phenomena". Thus, the triangle between spatial and analogic thinking and geospatial technologies promotes a constructivist learning environment, where inquiry and collaborative learning are encouraged.

Methodological Approach. Design-Based Implementation Research

In the present case, the project carried out shows two teaching techniques, the geographical narrative workshop and fieldtrip, which aligns with the Design-Based Implementation Research. This methodological framework is based on the principles proposed by the Design-Based Research Collective (Baumgartner et al. 2013). McKenney and Revees (2018) emphasize that educational design research merges scientific inquiry with the systematic development and implementation of solutions to educational problems. This methodology was applied to the campaign project to spread ocean literacy (Garcia de la Vega 2021).

The principles behind this research method have four principles: joint problems of teaching and learning practice; iterative and collaborative design; development of research theory; and, finally, capacity to sustain change in systems (Fishman et al. 2013). The design-based implementation research model offers insights into educational complexity and poses a challenge for the implementation of productive curricular adaptations (Choppin et al. 2013). Productive adaptations constitute a kind of advanced teacher professional practice, which requires extensive knowledge and instructional skills DeBarger et al. (2013). These authors call for new models for teacher professional development.

Design-based research has recently been described as a potentially fruitful methodology for generating causal accounts of learning and instruction that could form the basis for systematic. Design-based research methods focus on designing and exploring the whole range of designed innovations. Thus, didactic resources, activities strategies and plannings and curricula. This methodology goes beyond designing and testing interventions, because the educational interventions embody theoretical assertions about learning process and reflects a commitment to understanding the relationships between educational theory, planning and didactic resources (Baumgartner et al. 2013). At the same time, research on specific interventions can contribute to theories of learning and teaching. Consequently, research on the design of educational interventions could contribute to supporting and building other learning theories.

Scott et al. (2020) state that these "learning ecologies" are grounded in theories of learning, produce measurable changes in student learning, generate design principles that guide the development of instructional tools, and are enacted using extended, iterative teaching experiments. There are also approaches from engineering (Jung and Mercier 2023), which try to validate their learning tools with this method. Precisely, this educational research methodology comes from engineering and design researchers. Thus, Brown (1992) stated that, as a design scientist, I attempt to engineer innovative educational environments and simultaneously conduct experimental studies of those innovations. This involves orchestrating all aspects of a period of daily life in classrooms.

However, the type of educational research based on design has also reached geography. In one case, four projects are analysed using this methodology (Feulner et al. 2021), in another case, it is about applying the methodology (Garcia de la Vega 2021). And, finally, it is shown how the described characteristics of Design-based Research can guide future research projects and, thus contribute to overcoming the theory–practice gap in the context of educational research and school practice (Feulner et al. 2021).

Geographical Workshop and Fieldtrip

This section addresses the two teaching techniques used in the design based on educational research, geographical workshop and fieldtrip. The workshop presents three phases: exposition, experimentation and expression (Garcia de la Vega 2022b). The fieldtrip has a solid didactic approach in the teaching of geography. Fuller et al (2006), Herrick (2010) and Hammond (2018) show different arguments about the fieldtrip as a boost to conceptual geographic knowledge. The fieldtrip also assesses disciplinary competencies (Garcia de la Vega 2022a) and the fieldtrip lies its interest in the strength of the place-based learning approach, together with the GIS instrumentation and resources (Langran and DeWitt 2020).

The first phase, the exposition, consisted of an inquiry learning process through searching for and selecting valuable information and reading maps and documents. The second phase was the fieldtrip itself, where the documentary and cartographic information was contextualized. This phase corresponded to experiential and emotional learning in contact with the students' immediate environment and heritage landmarks. Finally, the expression phase consisted of the spatial representation of the coast with all the geographic and heritage elements, identified in the cartography and experienced in the fieldtrip. This representation was made with sand on the beach in a large group, clay in the classroom in groups of five students and, individually, it was also carried out with paint.

Preparation for the fieldtrip in the classroom using digital tools and in the fieldtrip itself, traditional geospatial resources have been used, such as the compass. The first consisted of reviewing geographical, historical and heritage content, through the use of digital resources. This required students to be able to select information to learn about the watchtowers and their historical and heritage value. In addition, they had to look for geographical terms related to the coastal topography and the wealth of the region. After the fieldtrip, the second phase of the workshop was held, where the narrative or artistic expression of the heritage milestones in their geographical context was carried out (Fig. 3 and 4).

The first and second phases of the geographical workshop were held before the fieldtrip and ended with the third phase in the classroom. Thus, the first phase consisted of an exposition, which was diluted in a dialogical discourse with the students. In this discourse, the heritage landmarks and the technological resources to look for them on their tablets were presented. The students then conducted research into the date of construction of the watchtowers, recognizing the agricultural, fishing and mining wealth of the historical and present-day region.

Thus, in these searches, they had to perform curatorial tasks, identifying and selecting rigorous information, always with subtle help of the teachers. In the second part, individually, through a drawing, or in small groups of students, they represented

Fig. 3 Exposition phase at workshop with K-12 in classroom. Students are curator and cartography readers with help of teachers. *Source* Photograph taken by the chapter's authors

Fig. 4 Experimentation phase at workshop with K-12 in classroom. Students are curator and cartography readers with help of teachers. *Source* Photographs taken by the chapter's authors

their geographical impressions of the coast. Here, they tried to apply their own knowledge, as well as the learnings of that day. They tried to find spatial relations between the natural landmarks, representing the hills and cliffs and beaches. And they tried to feel out a scale between the different geographical elements to give meaning to the representation of molding in clay or in the sand of the beach.

On the fieldtrip, visits to the surroundings of the watchtowers, to enable them to activate their spatial thinking skills. These skills consisted of locating heritage landmarks and linking them to geographical features. Thus, since in the classroom, the location of the landmarks had been reviewed on a cartographic viewer. This was the time to check if the students were able to orient themselves with the compass. Thus, skills are applied to cartography and navigation. They also established distances, so they were able to apply it in the representation of sand on the beach or clay in the classroom. Because everyone had to contribute their view of the landscape, making a part of it to later fit it into the joint construction. During the fieldtrip, the students were able to recreate the situations of that time. If an attack was coming, they would set fire, light at night or smoke during the day, which would allow them to warn of the attack to the inhabitants.

Among these types of reasoning, it is convenient to individuate the ones of a spatial nature. To do so, it needs to be identified that the spatial reasoning aimed at performing tasks of planning, supervision and assessment that are closely related to the spatial capacities. Analogical thinking is a mental process that consists of being confronted with novel situations by applying prior experiences learnt, according to a similar reasoning structure (Holyoak and Thagard 1997) (Figs. 5, 6 and 7).

Fig. 5 Torre de los Caballos. *Source* A. Garcia de la Vega, 2024

Fig. 6 Torre de Santa Isabel. *Source* A. Garcia de la Vega, 2024.

Fig. 7 Torre de Santa Elena. *Source* A. Garcia de la Vega, 2024.

Results, Discussion and Findings

The project was carried out with 83 primary school students from two schools in the province of Murcia (Spain). These students were grouped according to their educational level and age (10 and 11 years). The groups have been designated as follows: G1 with 24 students aged 11, G2 (18 students aged 11), G3 (20 students aged 11) and G4 (21 students aged 10). There are a small number of Maghrebian and Latin American students (two in each case per class). These students know the same landscapes as the rest of the students due to the permanent residence of their parents. However, students of European origin (British and Ukrainian) must be added, due to a temporary and/or recent residence of their parents. The two schools are public and are located in coastal municipalities of Murcia. The proximity to the coast of these schools is a didactic boost to meaningful learning. This educational assumption has been confirmed when the students show their links with natural places (hills, beaches) and watchtowers.

Results

The project was carried out through different dynamics inside and outside the classroom. In the first phase of the workshop, dedicated to the search for information about watchtowers and the reading of online maps, pairs of students were formed. Usually, this type of small group always offers good results, especially since the teachers try to find complementary personalities. In addition, in this phase, the teachers maintained a dialogue with the students to promote emotion and enthusiasm and focus of interest to avoid dispersion of attention. In the second phase, in the fieldtrip, the large group was divided into two groups of students . Thus, each group was identified with the color of the handkerchief they had been assigned in the distribution Berber pirates or inhabitants of the coast. The handkerchief was an element that evoked a different dynamic than usual in the classroom and generated a certain empathy among the members of each group. There were no rewards, it was just a matter of responding with spatial strategies to possible entry points into the bay, interests in the agricultural, fishing and mining production of the region.

The spatial representations of the natural features and heritage landmarks of the coast were made in sand and clay in the classroom. On the beach, the group of students was subdivided into two large groups to represent the coast in three dimensions. In turn, these large groups were subdivided into groups of three or four students who tried to represent the coast landscape around one of the watchtowers. And, later, all the geographic areas were brought together. Finally, a student presented to the large group of the students the spatial representation made by the little group. Beach sand is very attractive to mold in K-12 education. However, it lacks sufficient cohesion to generate a small structure that resembles a hill or a watchtower. The limited possibilities of the sand were therefore compensated for by the participation of a very large group. In addition, the students had to build a coastline that was evocative in its possible details of coastal areas and watchtowers. This happened with some geographical sectors well-known to the students, such as a beach or an island. However, it was imperceptible for the details of the locations of the watchtowers and the geometric shape of the floor plant of the watchtower, square or circular plan, as instance.

In the classroom, the spatial representation in clay was carried out by groups of four students. Each student was responsible for sculpting in clay the details of a sector of the coast, where there were landscape features (natural and heritage). Here, the results were directly related to the relationship of the students with each other and due to their personality. Thus, there were students who wanted to represent "their watchtower" without any relationship with the rest of the work group. On the contrary, there were brilliant groups, a good harmony was generated and there was always a leader. The student, as leader, was able to combine the spatial representations of the other members of the group. In addition, the clay provided greater representation details and coupled all the landscape features.

And finally, the beach groups were able to make a spatial representation on paper, individually. In this case, there were different ways of representing the coast and the natural features and heritage landmarks. In all cases, the response to these free spatial representations generated emotion and enthusiasm in many students. This contrasted with many other students who were slow to react to know how to make the spatial representation. This points to a need to address free cartographic expression at an early age to progress towards mastery of the spatial skills.

To synthesize the database, three spatial dimensions have been established, two individual and one collective. The two individual dimensions are: recognition and elaboration of spatial thinking. The individual dimension of recognition refers to five competencies: Identifying and classifying the landscape feature; ascribing meaning to the shape; recognizing a shape of a landscape and outlining a pattern of landscape feature. The individual dimension of spatial thinking includes these competences: Drawing the shape of a landscape feature linked to heritage landmarks; Sketching the position of watchtower on the landscape; Tracing the orientation of watchtower on the landscape and pinpointing the location of watchtower on the landscape. The collective dimension corresponds to the scope of the activities carried out in small or large groups. The group competences are: Describing the shape of a landscape feature; Visualizing three-dimensional landscape features and reproducing three-dimensional landscape features. The first competence refers to describing the shape of a landscape feature and was carried out on the beach with sand to recognize the profile of the coast. The second competence is Visualizing and Reproducing three-dimensional landscape features, which is related to the representation of beach sand in large groups and white clay in small groups.

The mental maps were made at the end of the morning, after the cartographic session in the classroom and the fieldtrip. In general terms, there are differences between groups 1 and 4 compared to groups 2 and 3, since they belong to different schools in different locations. That is, groups 1 and 4 belong to a school located in a coastal town with a small population, while groups 2 and 3 belong to a school located in a town with a larger population and larger groups of students. The different origins of some students imply a reduced knowledge of the natural and cultural environment and an added difficulty in understanding the language. This was an added constraint for the second group of students to carry out the activity, who were assisted by the teachers.

The guidelines for making the mental maps were reduced to tracing the shape of the coast and placing the watchtowers in their places on a sheet of paper. They were also asked to try to orient the coast and distribute and place the watchtower in relation to the other watchtowers and features of the landscape (lighthouse, hill, town, boulevard and cliff). Nor was a guideline established regarding the position of the paper and the student, which led to four cases with the representation defined by a mirror effect. That is, most of the students made a representation oriented according to the cartographic consultations in class and the panoramic views provided by the extreme watchtowers of the gulf. These guidelines attempted to assess the acquisition of spatial concepts related to the location of the watchtowers, orientation of the main landscape features, the gulf, the spatial distribution of geographic and heritage features and the proposal of a scale, among others.

In general, a dominance of Cartesian representation with some artistic elements has been observed. This Cartesian representation may be due to the online cartographic consultation session. These artistic elements refer to a way of representing the features studied, both landscape and heritage. The inclusion of these artistic features is interpreted in two ways, the first refers to offering a more personal and experiential mental map; the second is related to a way of replacing the shortcomings of making a Cartesian cartographic representation. However, there are some students, close and far from the synthetic representation of the gulf, who add some childish elements foreign to the cartographic representation).

The results show a mastery of the geographical concepts (beach, coast, bay, gulf…) in terms of their Cartesian representation (competences 1 and 2), after a hands-on experience through the fieldtrip in the groups G1 and G4 (Fig. 8). There are students who tried to accurately draw the shape of the coast, adding all the natural and heritage features (Fig. 8a, b). In these two mental maps, two of the categories are recognized, following Grossner and Janelle (2014), position and spatial relation. Thus, in the first category, the mental maps show the acquired spatial concepts of position, distance, direction and orientation. As for the second category, proximity and distribution between the landmarks studied are recognized. All of them recognized the concept of the coast, except for two students, who chose a comic form to highlight the heritage value of the watchtower (Figs. 8c, e). The geographical terms presented in the session were the concepts of coastal relief presented in the curriculum. In addition, some terms were added that arose in the dialogue held with the students in class, as instance, inlet and gulf (Fig. 8b). These terms were in the cartography that they consulted, that appear with drawings in their concave shape (Fig. 8f).

In the groups G2 and G3, the students show the concave profile of the Gulf of Mazarron, trying to make a Cartesian representation. This feature in the drawn mental map requires a greater degree of synthesis, or rather, an attempt to approximate the representation of the read map. That is, trying to find lines and points that allow

Fig. 8 Mental maps with cartesian outline and artistic elements

the heritage landmarks to be located and oriented. However, as in Fig. 8, groups G1 and G4 represent the heritage landmarks with artistic features that even allow the geometric shape of the towers to be recognized. They also tried to recognize in the cartography the watchtowers and the natural features of the known coastal landscape, such as the hill located next to the school or the lighthouse. In two cases, the students tried to represent the contour lines (Figs. 9a, b, c and e). In any case, after the presentation and reinforcement of showing the shape of the geographic concept "gulf" in the cartographic session in the classroom and in the fieldtrip, it is striking that there are twenty students who do not manage to represent the shape of this concept. Even more so, because they are students who live next to the coast. However, in two groups (G-2 and G-3) they clearly recognized this concept, this could be due to being presented in class the weeks prior to the activity. The degree of precision in the location, distribution, and orientation of natural and heritage landmarks is similar to that of Groups G1 and G4, and even in the creative inclusion of these landmarks (Fig. 8). However, in these two groups, G2 and G3, two cases are observed that are different from the rest of the students (Figs. 9d,f). In the first mental map, the coasts have angular layouts, although the concave shape has been recognized (Fig. 9d). However, the heritage landmarks are confused with the digital resources consulted (cartographic viewer, Iberpix). In the second, the mental map resembles a simple sketch of the heritage landmarks, without names and similar to Fig. 8c,e of the other student groups (Fig. 9f).

There are three spatial competences, where the shape of the coast and the watchtowers are related, through position, orientation and location. The result of the position of the four watchtowers is better resolved than the other two competitions. The position of the watchtowers has been correct in 73/83 cases, considering the possibility of placing two watchtowers to validate the result. This result is better than the orientation (50/83 cases) and location (54/83 cases) of the watchtowers in the gulf, which requires a distribution in the landscape and a scale proposal. This is because the representations were freer, although the position between the watchtowers lacks the proper orientation (Figs. 8c,d, 9d,f). Or the representations lack the location in relation to the natural landmarks of the landscape, the gulf (Figs. 8c,d,e, and 9f).

Essentially, there were two styles of spatial representation, which could be defined as cartesian and artistic. Under the cartesian style, there is a larger group of students who represented precise lines, sometimes not, of the coast and its natural landmarks (hills, cliffs, beaches). The cartesian character could correspond to this need to replicate what they have seen in the first phase of the workshop. Now, almost all of them include some feature that broke this orthodox cartographic representation with the inclusion of figurative heritage landmarks. Thus, for example, the watchtower appeared drawn as it could be experienced in the fieldtrip, not as a point as it appears on the map. The most obvious case corresponds to the fact that two students represented the contour lines (As instance, Fig. 8b).

Discussion

The aim of this chapter is to promote spatial thinking by exploring the use of technologies. The coastal geographic sector, where the watchtowers are located, and the proximity of the school to this environment generate an intense emotional bond. Many families have visited the watchtowers, the students know the environment of the fieldtrip and even the father of one of the children works in the tuna fishing factory (*almadraba*). However, the schools have sufficient means to start GIS, although the teachers are unaware of the simplest applications and their implementation in the

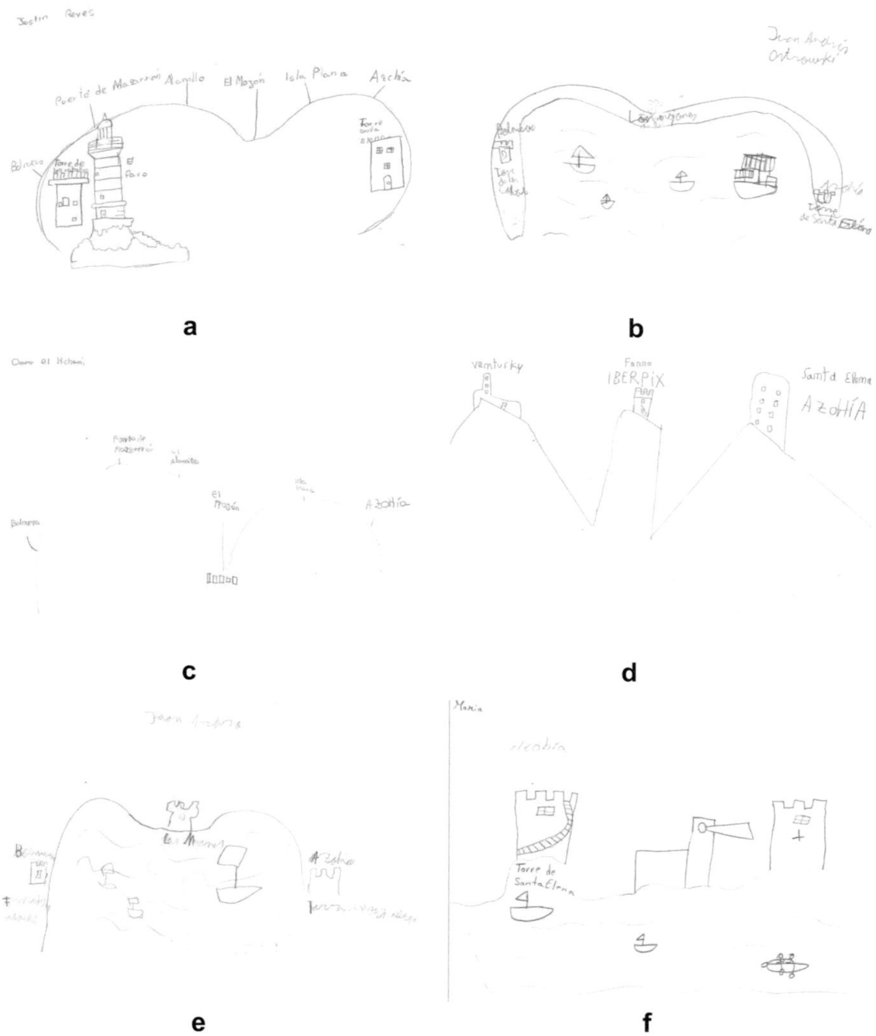

Fig. 9 Mental maps with cartesian outline and artistic elements

lesson plan. The use of geospatial technology in spatial skills and knowledge of spatial relationships demonstrates a link and a boost in learning between spatial and geographical thinking, aligned with Metoyer and Bednarz (2017).

This work is aligned with Steger et al. (2021), who consider the importance of children's emotional attachments to places as well as the centrality of human experiences in GIS. Perspective taking is the ability to imagine points of view other than one's own (Brucato et al. 2024). In relation to spatial perspective taking, these authors have not found associations with social perspective taking (cognitive or affective). Consequently, these authors suggest that the pattern has support for a dissociative model in which spatial perspective taking is a distinct cognitive construct. Catling (2022) highlights ten big ideas from geography, linked to geographical thinking. However, this author aims to apply the ideas to the world of childhood, at its scale and in its setting. Therefore, spatial experiences suggest an intrinsic and personal value. This drives emotion and experience in relation to personal exploration of space and, therefore, its graphic representation.

However, we consider the suggestions made by teachers by Zwartjes and De Lázaro (2019) to be important. In this document, it is suggested: 1. Improve teachers' skills on visualizing geographical information on maps; 2. Use GI Science Methodology for selected topics; and 3. Find topics in your curriculum that fit GI-Learner contents and take advantage of ready-to-use. Thus, we must relate these competence suggestions from teachers with the spatial concepts that students must acquire. According to Michel and Hof (2013), three spatial thinking dimensions are distinguished. These authors relate these dimensions with levels of acquisition of geospatial skills: spatial concepts, spatial representation and spatial reasoning.

If applied to this project, according to the limitations of schools and teachers, only spatial concepts and spatial representation could be developed. Gersmehl (2023) highlights the eight spatial reasoning, which are difficult to achieve in K-12, since they require a higher range of complexity (comparison, proximity, transition…). Of these three dimensions, in this research the spatial concepts have been addressed to explore the possibilities of developing analogical thinking. By promoting this type of thinking through activities with the use of online maps, spatial thinking was promoted in students. Teacher training in technological skills has also been promoted, as teachers recognized the shortcomings related to access to GIS.

Among the ten categories of spatial thinking established by Grossner and Janelle (2014), five were addressed in this research, as well as some of the related concepts: Context (situation, global, local); Position (position, distance, direction, orientation); Spatial relations (proximity, distribution, spatial hierarchy); Spatial transformations (scale, rotation) and Representation (graph, cognitive map). These selected categories were addressed both in group work and in individual work. That is, the categories were addressed in the classroom, so that "context" and "position" could be offered to the heritage milestones in relation to the landscape. Spatial relations and spatial transformation were applied both in the classroom (scale) and on the fieldtrip (orientation, proximity). The representations of the coast with its heritage milestones, made in the sand on the beach and in white clay in groups, required the categories of position, spatial relations and representation. While mental maps collect only a few spatial concepts, this means that in the group dynamics, whether in the classroom or outside it, through online cartography and satellite images and teaching fieldtrips, a greater number of spatial categories were reviewed. However, not all the spatial concepts suggested by Grossner and Janelle (2014) were addressed in the spatial thinking categories. Given the conditions of the technological resources of the schools and the geo-competencies of the teachers, expressed above, the students' geoskills were explored. To do so, the proposal of National Research Council (2006) has been used, adapted and applied to this research, as a set of geospatial concepts and skills to assess spatial competencies (Table 1). The hierarchy of complexity for spatial concepts in

Table 1 Categories of spatial thinking

Categories of spatial thinking	G1-24	G2-18	G3-20	G4-21	Total-83
1. Identifying a landscape feature by its shape (coast)	22	18	20	21	81
2. Classifying a landscape feature by its shape (coast)	21	18	20	17	76
3. Ascribing meaning to the shape of landscape features (bay, gulf)	14	18	20	11	63
4. Recognizing a shape of landscape feature linked to heritage milestones	20	15	17	17	69
5. Outlining a pattern of landscape features linked to heritage milestones	18	15	16	14	63
6. Drawing the shape of a landscape feature (bay, gulf)	14	13	15	6	48
7. Sketching the position of heritage milestones on the landscape	20	18	19	16	73
8. Tracing the orientation of heritage milestones on the landscape	14	18	17	9	50
9. Pinpointing the location of heritage milestones on the landscape	13	15	16	13	54
10. Describing the shape of a landscape feature					
11. Visualizing 3D landscape features					
12. Reproducing 3D landscape features					

Source Adapted to this project and modified from National Research Council (2006)
Landscape Features linked to Heritage Milestones

geography has also been considered (Jo and Bednarz 2009). These authors identified three levels: primitive, simple spatial, and complex-spatial, respectively, the concepts may be location, distance, and distribution. In this sense, only the primitive level is identified and assessed.

Findings

In the two main actions of the project, the geographic workshop and the fieldtrip, there are relevant findings in relation to spatial thinking. These findings are also related to an intuitive exploration of online cartography and the documentation offered in web search engines. On the one hand, in the first phase of the workshop, where the students had the role of curator, the value of working in pairs in achieving the objectives has been shown. Thus, the students showed the acquisition of some cartographic issues, related to the symbology of the legend. In fact, toponymy, contour lines, altitudes, population, towns were recognized by the students. Recognition of those places close to their home, school and places visited was also observed, for example, such as the hill located behind the schools (*El Cabezo del Horno, 285m*).

In the fieldtrip, the students showed expectation to recognize their known space. Thus, enthusiasm appeared in recognizing known spaces in their environment. During the fieldtrip, students demonstrate their spatial and analogical thinking skills. After the workshop session, students put spatial skills into practice in a real context, with express reference to local heritage. A cultural heritage that students also value and contextualize the content. Content that, in its relationship with spatial thinking, involves navigation. The accessibility of mobile devices and freely accessible online cartography provide good conditions for promoting these two types of thinking.

Mental maps reflect instrumental deficiencies both in students and in the curriculum. Teachers acknowledged their lack of training in GIS and, therefore, this is a drawback, since the schools have sufficient technological means (supports and web). This points to a need for teacher training in GIS applied to interdisciplinary teaching approaches, such as relating geographical concepts and heritage features. Even more so, since teachers highly valued outdoor activities with a presentation of indoor content.

Conclusion

Throughout this chapter, it has been proven that promoting a project that brings together heritage content is able to develop spatial and analogical thinking. Geographical, historical and heritage content promotes cartographic reading and navigation. At the same time, the project reinforces curriculum content and promotes a wide range of cognitive abilities, such as analogical thinking. Cartographic reading, together with searching for content on a heritage topic, promotes meaningful learning. Cartographic reading with intuitive access to mobile device applications stimulates more complex cognitive processes from inquiry learning. Thus, carrying out the fieldtrip allows for experiential learning from proximity to places of heritage interest to link it to spatial thinking. The fieldtrip allows for contextualizing heritage knowledge and relating it to navigation and cartographic reading. Direct experiences lived in open environments, contextualized in real life, are exciting because they challenge reasoning. Similarly, alternating applications (map viewers, programs, web search engines) on different mobile devices and computers, as well as analog measuring instruments (compass and paper map) enhance the exploratory capacity of spatial thinking. Location, direction, and distance allow citizens to develop autonomy and strengthen the formation of spatial and analogical thinking. In this way, education fosters a citizenry committed to planning the future, integrating geospatial technologies.

In summary, this chapter shows the interrelation between curricular contents linked to heritage and spatial and analogical thinking skills. It is shown that teachers could foster spatial thinking in students (K-12) through simple teaching strategies that are preparatory to the use of geospatial technologies. This helps prepare students to apply these geospatial skills in the environment and in their real context. Experiential experience and proximity to heritage are key to getting students excited about learning simple spatial skills through fieldtrips. Consequently, this prepares future citizens through instruction on the autonomous management of everyday spatial situations or even how to propose effective and efficient solutions in professional environments. Finally, the experiential and experimental orientation of the chapter reinforces a complementary use of geospatial skills and the use of geospatial technologies. Perhaps, in this educational research design, a greater application of geospatial technologies has been left pending. However, this has been conditioned by the technological characteristics of each school and, essentially, and higher acquisition of technological competences in teachers. If there were a higher technological level in the schools, collaborative cartography could be carried out.

Acknowledgements To the Ministry of Defense of Spain for granting aid to project 130-129279. To the effective collaboration in the project of Juan Luis Arceda and the participation of the school management, as well as that of teachers and students.

References

Bednarz SW, Jo I, Shin E (2022) Spatial thinking in primary geography. In: Kidman G, Schmeinck D (eds) Teaching primary geography. Springer, Cham (SW), pp 133–144

Brown AL (1992) Design experiments: Theoretical and methodological challenges in creating complex interventions in classroom settings. J Learn Sci 2(2):141–178

Brucato M, Newcombe NS, Chein J (2024). Spatial perspective taking is distinct from cognitive and affective perspective taking. bioRxiv 2024.06.16.599248; https://doi.org/10.1101/2024.06.16.599248d

Catling S (2022) The challenge of geography for primary children. Zeitschrift für Geographiedidaktik 49. https://doi.org/10.18452/25364

Choppin J, Debarger A, Beauvineau Y, Moorthy S (2013) Designing for productive adaptations of curriculum interventions. In Penuel W, Cheng BH, Fishman BJ, Allen AR (eds) Design-based implementation research: theories, methods, and exemplars. yearbook of the National Society for the Study of Education. Teachers College, New York, pp 298–319

DeBarger AH, Choppin J, Beauvineau Y, Moorthy S (2013) Designing for productive adaptations of curriculum interventions. In: National Society for the Study of Educational yearbook, 112 (2): 443–463

Baumgartner E, Bell P, Brophy S, Hoadley Ch, Hsi Sh, Joseph D, Orrill Ch, Puntambekar S, Sandoval W, Tabak I (2013) Design-based research: an emerging paradigm for educational inquiry. Educ Res 32(1):5–8. https://doi.org/10.3102/0013189X032001005

Feulner B, Hiller J, Serwene P (2021) Design-based research in der Geographiedidaktik–Kernelemente, Verlaufsmodell und forschungsmethodolo gische Besonderheiten anhand vier ausgewählter Forschungsprojekte. EDeR–Educ Des Res 5(2):1–32. https://doi.org/10.15460/eder.5.2.1576

Fishman B, Penuel WR, Allen A, Cheng BH, Sabelli N (2013) Design-based implementation research: an emerging model for transforming the relationship of research and practice. In: Fishman P, Allen, Cheng (eds) Design-based implementation research: theories, methods, and exemplars. National Society for the Study of Education yearbook, vol 112, issue 2. Teachers College Record, New York, pp 136–156

Fuller I, Edmondson S, France D, Higgitt D, Ratinen I (2006) International perspectives on the effectiveness of geography fieldwork for learning. J Geogr High Educ 30(1):89–101

Garcia de la Vega A (2019) Spatial thinking ability acquisition through geospatial technologies for lifelong learning. In: De Miguel González R, Donert K, Koutsopoulos KC (eds) Geospatial technologies in geography education. Springer, Cham (SW), pp 21–40
Garcia de la Vega A (2021) Design-based implementation research for exploring the ocean. A geographical perspective. In: Koutsopoulos KC, Stel JH (eds) Ocean literacy. Understanding the Ocean. Springer, Cham (SW), pp 115–148
Garcia de la Vega A (2022a) A proposal for geography competence assessment in geography fieldtrips for sustainable education. Sustainability 14(3):1429. https://doi.org/10.3390/su14031429
Garcia de la Vega A (ed) (2022b) Taller de narrativa científica sobre el cambio climático. Octaedro, Barcelona
Gersmehl PJ (1991) The language of the maps. Indiana University of Pennsylvania and National Council for Geographic Education, Indiana
Gersmehl PJ (2014) Teaching geography. Guilford Press, New York
Gersmelh PJ (2023) Brain science and geographic thinking: a review and research agenda for K-3 geography. Educ Sci 13:1199. https://doi.org/10.3390/educsci13121199
Goodchild MF, Janelle DG (2010) Toward critical spatial thinking in the social sciences and humanities. GeoJournal 75:3–13. https://doi.org/10.1007/s10708-010-9340-3
Grossner K, Janelle DG (2014) Concepts and principles for spatial literacy. In: Montello DR, Grossner K, Janelle DG (eds) Space in mind: concepts for spatial learning and education. MIT Press, Cambridge, MA
Hammond L (2018) The place of fieldwork in geography education. In: Jones M, Lambert D (ed) Debates in geography education. Routledge, Oxon
Herrick C (2010) Lost in the field: ensuring student learning in the 'threatened' geography fieldtrip. Area 42(1):108–116
Holyoak KJ (1995) Problem solving. In: Smith EE, Osherson DN (eds) An invitation on cognitive science, vol 3, 2nd edn. MIT Press, Cambridge, MA, pp 267–296
Holyoak KJ, Thagard P (1997) The analogical mind. American Psycolo-Gist 52(1):35–44
Jekel T, Gryl I, Schulze U (2015) Education for spatial citizenship. In: Osvaldo Muñiz Solari O, Demirci A, van der Schee J (eds) Geospatial technolgies and geography education in a changing world. Geospatial practices and lessons learned. Springer, Tokyo, Japan, pp 35–52
Jo I, Bednarz SW (2009) Evaluating geography textbook questions from a spatial perspective: using concepts of space, tools of representation, and cognitive processes to evaluate spatiality. J Geogr 108(1):4–13. https://doi.org/10.1080/00221340902758401
Jung J, Mercier E (2023) Design-based implementation research: milestones and trade-offs in designing a collaborative representation tool for engineering classrooms. Educ Technol Res Dev 71:2457–2481. https://doi.org/10.1007/s11423-023-10288-z
Langran E, DeWitt J (2020) Navigating place-based learning. Springer
McKenney S, Reeves TC (2018) Conducting educational design research. Routledge, London
Metoyer SK, Bednarz SW, Bednarz RS (2015) Spatial thinking in education: concepts, development, and assessment. In: Osvaldo Muñiz Solari O, Demirci A, van der Schee J (eds) Geospatial technologies and geography education in a changing world. Geospatial practices and lessons learned. Springer, Tokyo, Japan, pp 21–34
Metoyer S, Bednarz R (2017) Spatial thinking assists geographic thinking: evidence from a study exploring the effects of geospatial technology. J Geogr 116(1):20–33. https://doi.org/10.1080/00221341.2016.1175495
Michel E, Hof A (2013) Promoting Spatial Thinking and Learning with Mobile Field Trips and eGeo-Riddles. GI_Forum 2013: Creating the GISociety (Eds. Jekel, T., Car, A., Strobl, J., Griesebner, G.), Wichmann Verlag Berlin, 378–387. https://doi.org/10.1553/giscience2013s378
National Research Council (2006) Learning to think spatially. The National Academies Press, Washington, DC. https://doi.org/10.17226/11019
Newcombe N (2016) Thinking spatially in the science classroom. Curr Opin Behav Sci 10:1–6. https://doi.org/10.1016/j.cobeha.2016.04.010

Scott EE, Wenderoth MP, Doherty JH (2020) Design-based research: a methodology to extend and enrich biology education research. Cross-Disciplinary Res Biol Educ Essay 19(es11):1–12. https://doi.org/10.1187/cbe.19-11-0245

Sinton DS, Kolvoord B, Gersmehl P, Bednarz S, Uttal D (2013) The people's guide to spatial thinking. National Council for Geographic Education, Washington, DC

Sinton DS (2016) Spatial thinking and GIS. In: Burte H, Kauppinen T, Hegarty M (eds) Proceedings of the workshop on teaching spatial thinking from interdisciplinary perspectives (TSTIP 2015) at COSIT 2015 in Santa Fe, NM

Steger A, Evans E, Wee B (2021) Emotional cartography as a window into children's well-being: visualizing the felt geographies of place. Emotion Space Soc 39:100772. https://doi.org/10.1016/j.emospa.2021.100772

Uttal DH, Meadow NG, Tipton E, Hand LL, Alden AR, Warren C, Newcombe NS (2013) The malleability of spatial skills: a meta-analysis of training studies. Psychol Bull 139(2):352–402. https://doi.org/10.1037/a0028446

Wilson BG (ed) (1996) Constructivist learning environments: case studies in instructional design. Educational Technology Publications, Englewood Cliffs, NJ

Zwartjes L, de Lázaro y Torres ML (2019) Geospatial thinking learning lines in secondary education: The GI learner project. In: de Miguel González R, Donert K, Koutsopoulos K (eds) Geospatial technologies in geography education. Springer, Cham, pp 41–61

Collaborative Mapping of Endangered Cultural Heritage to Promote Sustainable Cities

María Sebastián López, Ondrej Kratochvíl, Juan Mar-Beguería, and Rafael De Miguel González

Abstract The preservation of cultural heritage is a crucial component of sustainable urban and territorial development. This study explores the role of collaborative mapping as both a pedagogical and civic tool for raising awareness about heritage conservation and promoting sustainable spatial citizenship. Using a mixed-methods approach, the study evaluates how students and local communities engage with geospatial technologies to document, assess, and propose solutions for at-risk heritage sites. The research is structured at two different scales: a national study across Spain and a localized case study in Zaragoza. The integration of collaborative mapping in educational settings not only enhances spatial and geographic competencies but also fosters a more critical and participatory approach to urban sustainability. This study demonstrates that digital geolocation tools can be leveraged for active heritage conservation, transforming citizens as agents of change. The results underscore the importance of interdisciplinary methodologies that connect geographic education, heritage studies, and civic engagement for the sustainable management of cultural assets.

Keywords Geography education · Geolocation · Heritage · Digital skills · Geographical skills

M. Sebastián López (✉) · O. Kratochvíl · J. Mar-Beguería · R. De Miguel González
Faculty of Education, ARGOS-IUCA Research Group, University of Zaragoza, Zaragoza, Spain
e-mail: msebas@unizar.es

O. Kratochvíl
e-mail: onkra@unizar.es

J. Mar-Beguería
e-mail: j.mar@unizar.es

R. De Miguel González
e-mail: rafaelmg@unizar.es

R. de Miguel González and P. Rivero Gracia (eds.), *Geospatial Technologies for Heritage Education*, Key Challenges in Geography,
https://doi.org/10.1007/978-3-031-96372-8_6

Introduction

Since time immemorial, humans have sought to understand their environment in a precise manner to control and manage various aspects of a given territory. This includes natural elements such as rivers, agricultural fields, etc., as well as man-made and social elements such as property boundaries, buildings, and associated cultural heritage (Stamenov and Stamenova 2021).

All these elements, integrated in the perceived landscape, share a common characteristic: their spatial component. This is why everything around us can be georeferenced and cartographically represented, making it possible to first acquire knowledge about it and then incorporate it into decision-making for territorial development. In essence, geoinformation or geographic information is both useful and necessary for understanding and managing a territory. The same applies to cultural heritage, as all types of heritage are linked to specific geographic locations—whether tangible assets, which are tied to a precise point on the map, or intangible assets, which are associated with broader geographical areas (De Miguel 2024).

Heritage knowledge is essential for effective territorial development and its management from a cultural perspective. This approach should be rooted in the local history and traditions of a place, particularly when the aim is to preserve cultural memory and identity.

Collaborative Mapping of Cultural Heritage: Towards the Achievement of SDG 11.4

Aware of this, the preservation of cultural heritage is one of the UN's priority Sustainable Development Goal (SDG) 11.4, adopted in 2015, that emphasizes the need to protect and safeguard the world's cultural and natural heritage. Rural and urban heritage, natural and cultural, plays a fundamental role in shaping identities, preserving collective memory, and fostering social cohesion through the landscape. It includes not only historical monuments and architectural landmarks but also traditional neighborhoods, markets, public spaces, and cultural practices that define the character of a place.

However, urban heritage faces significant challenges due to rapid urbanization, population growth, and economic pressures, giving precedence to pragmatic visions of the territory (Martinez de Pisón 2010). The expansion of modern infrastructure, real estate development, and inadequate urban planning often leads to the destruction or deterioration of historic sites and culturally significant areas. In many cities, gentrification and commercialization threaten the authenticity and accessibility of heritage spaces, displacing local communities and altering the social fabric of historic districts. Additionally, environmental factors such as pollution, climate change, and natural disasters further endanger urban heritage, accelerating its degradation.

Community involvement is a crucial element in the heritage conservation process. Local residents, cultural organizations, and heritage experts should actively participate in decision-making to ensure that preservation efforts align with the needs, values, and identities of the communities that inhabit these spaces. Without this engagement, heritage conservation risks becoming a top-down process disconnected from the lived experiences and priorities of local populations (Bokova 2023).

This paper argues that educational strategies based on collaborative mapping play a key role in fostering a sense of ownership and responsibility among citizens, encouraging them to take an active part in heritage preservation. By providing communities with tools for visualizing heritage, reporting its state of conservation, and advocating for its protection and enhancement, these initiatives promote a more engaged, participatory, and conscious citizenry. This, in turn, strengthens the collective commitment to building more sustainable cities, where cultural and historical assets are valued as integral components of urban identity and development.

Digital Competences on Cultural Heritage Learning and Awareness: Quality Education for Sustainable Cities and Communities (CIVITAS III)

In this sense, geographic education systems are emerging as the best tools because they can be used to acquire both transversal digital and specific geographic competencies and, in addition, to promote a reflective and sustainable spatial citizenship with its environment. Since the early 21st century, the rise of online Geographic Information Systems (GIS), web-based GIS viewers, and digital atlases has revolutionized the way spatial data is accessed, analyzed, and applied. This technological progress has profoundly impacted various fields, including heritage studies, where GIS has become an indispensable tool for documentation, analysis, and preservation. In recent years, research on the use of GIS in heritage conservation has expanded significantly, highlighting its potential for supporting the sustainable management and spatial planning of cultural and historical assets.

At the same time, there has been an exponential growth in the development of educational resources aimed at integrating GIS into teaching and learning, particularly in the Social Sciences (Álvarez and Lázaro-Torres 2018; Buzo Sánchez et al. 2022, 2023; De Miguel et al. 2019; DeMiguel and Lázaro 2020; De Miguel and Sebastián 2022; Sebastián and De Miguel 2020; Sebastián et al. 2023; Serrano and Sebastián 2024). These resources are designed to enhance transversal, and in particular digital competencies by fostering spatial literacy, critical thinking, and data analysis skills. One of the most effective approaches in this context is collaborative mapping, which enables students, educators, and researchers to engage in participatory heritage documentation, promoting a more inclusive and community-driven approach to its conservation (Giovene di Girasole et al. 2019).

In particular, GIS and collaborative mapping permit to educators and learners to go beyond traditional heritage analysis methods, actively engaging with real-world challenges related to cultural preservation, urban development, and environmental sustainability (Amichai-Hamburger 2008). This approach not only provides to students with some valuable digital and analytical skills but also empowers them to become active participants in heritage protection and spatial justice initiatives.

One example of these strategies is the CIVITAS III project, which has among its primary objectives the analysis of how school communities acquire digital geographic competencies through the use of collaborative mapping. This methodology has been implemented in both lower secondary education students (*Educación Secundaria Obligatoria*) and teacher training students of bachelor's and master's degrees. These teacher education programs are focused on Early Childhood, Primary, and Secondary Education across three Spanish Autonomous Communities: Aragón, Andalucía, and Castilla y León.

Specifically, an ArcGIS Survey123 app has been developed to facilitate an interactive and systematic assessment of cultural heritage sites (Fig. 1), incorporating the nomenclature and analytical guidelines for heritage proposed by UNESCO (2021, 2023). This tool enables participants to:

(1) Identify and geolocate the heritage asset—Users can pinpoint the exact location of a heritage site, ensuring accurate spatial documentation. In addition, in the case of tangible heritage, it allows them to add a photograph that identifies them.
(2) Evaluate its significance from a personal and identity-based perspective—Participants reflect on the cultural, historical, and emotional value of the site, linking it to local and collective identity.
(3) Assess its state of conservation—The app allows for a structured evaluation of the site's physical condition, identifying signs of deterioration, neglect, or potential threats.
(4) Determine the type of risk it faces—Users classify risks, including environmental hazards (e.g., climate change, pollution), human-induced threats (e.g., vandalism, urban expansion), or structural decay.

By integrating geospatial technology with heritage education, this app promotes active engagement, fosters spatial awareness, and supports data-driven decision-making in cultural heritage conservation.

Objectives

Based on the aforementioned ideas, this study pursues two main objectives, which have been addressed using different methodologies and spatial analyses:

(1) Identify the types of heritage that are considered most relevant or most easily recognizable by the general population involved in the study (tangible or intangible) and determine which types of threats are most easily identifiable

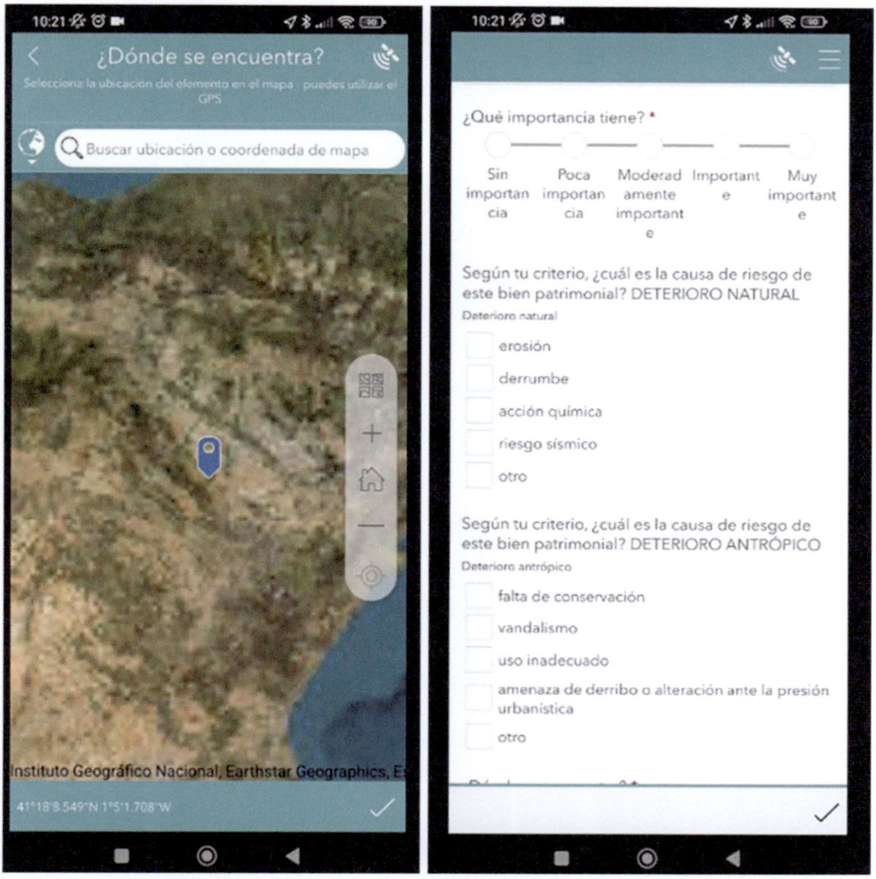

Fig. 1 Screenshots of the ArcGIS Survey123 mobile application and the integrated questionnaire, used for collaborative mapping of endangered cultural heritage

(anthropic or natural). This objective seeks to understand the perception of cultural heritage within communities. Additionally, it examines the extent to which different heritage elements are perceived as endangered, whether due to human actions (urban expansion, pollution, neglect) or natural causes (climate change, erosion, natural disasters, risks, and hazards).

(2) Assess how collaborative mapping strategies foster a more active, critical, and reflective citizenship that contributes to sustainable cities. This objective explores the role of participatory geospatial technologies in enhancing civic engagement and spatial awareness. It investigates how involving citizens in heritage documentation through digital mapping tools (e.g., GIS-based applications) promotes social responsibility, encourages heritage conservation, and integrates local knowledge into urban sustainability policies.

By addressing these objectives, this research aims to bridge the gap between heritage studies, geospatial analysis, and civic participation, emphasizing the importance of digital literacy in cultural heritage conservation and urban sustainability planning.

Research Approach and Analysis Scales

To efficiently address both objectives, the research has been divided into two levels of analysis:

(1) **A broad-scale analysis at the national scale (Spain)**—This approach examines the types of heritage mapped by pre-service teachers specializing in Early Childhood, Primary, and Secondary Education. Additionally, it provides the most commonly identified threats to heritage sites, offering insights into the general perception of cultural heritage and the factors contributing to its degradation across different regions of Spain.
(2) A detailed **case study in the city of Zaragoza (Aragon, Spain)**—This focused analysis explores how collaborative mapping of urban cultural heritage, combined with critical reflection on its state of conservation, can foster spatial awareness and promote an engaged, reflective, and sustainability-oriented citizenship. By assessing participants´ interpretation and documentation of heritage in their immediate surroundings, this study aims to highlight the role of digital mapping in shaping spatial literacy, civic responsibility, and sustainable urban development advocacy.

This multi-scale approach allows for both a broad assessment of national trends in heritage perception and conservation challenges, as well as a localized, in-depth exploration of how collaborative mapping methodologies influence citizen engagement and urban sustainability awareness.

Spanish Scale

At this spatial scale, the results obtained by participants in the research project Digital Competences, Learning Processes, and Cultural Heritage Awareness: Quality Education for Sustainable Cities and Communities (PID2020-115288RB-I00), funded by the Spanish State Research Agency and the Ministry of Science and Innovation, are analyzed.

This study does not focus on an educational perspective—already addressed by colleagues in this volume—but rather examines the findings derived from the mapping process conducted using ArcGIS Survey123. The analysis aims to evaluate how geospatial technologies contribute to the documentation, recognition, and assessment of cultural heritage, offering insights into the types of heritage identified and the perceived threats affecting them.

Field data collection was conducted across the entire Spanish territory from March 2023 to December 2024. A total of 1098 heritage records were obtained, with 91.98% of them corresponding to tangible heritage and only 8.02% to intangible heritage.

Geographically, while the mapping covers all of Spain, a concentration of records is observed in the Aragón and La Rioja (Fig. 2). This distribution suggests a regional bias in data collection, potentially influenced by the location of participating institutions.

As a result of the mapping, both quantitative and qualitative information was obtained for each georeferenced asset: (1) type of heritage (tangible/intangible); (2) degree of importance (likert scale); (3) personal assessment of the heritage asset's significance; (4) type of deterioration (natural/anthropic); (5) specification of causes of deterioration: natural (erosion, collapse, chemical action, seismic risk, other) or anthropic (lack of conservation, vandalism, inappropriate use, threat of demolition or alteration due to urban pressure, other).

All the information was analyzed through descriptive statistics using SPSS and a qualitative analysis with Atlas.Ti.

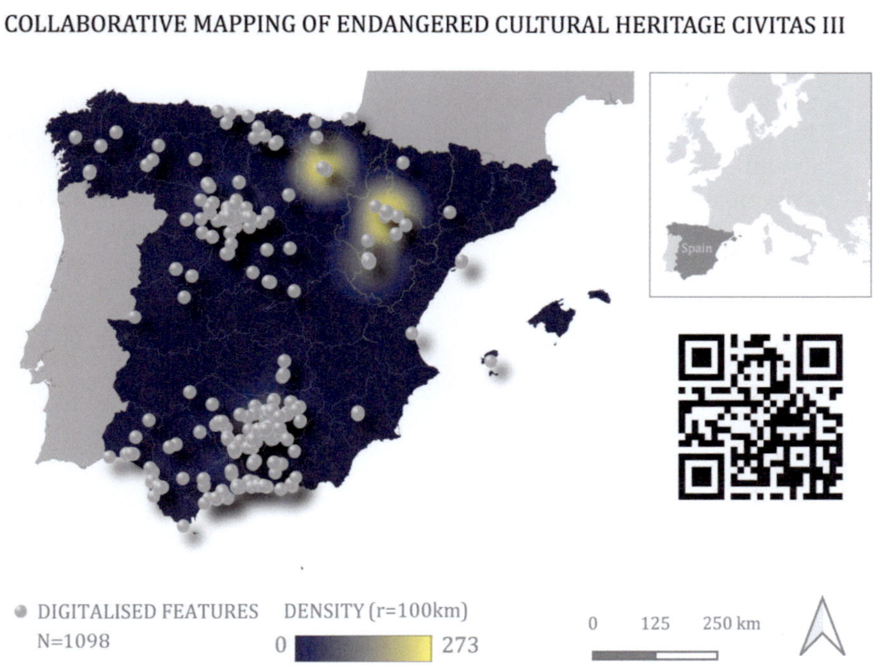

Fig. 2 Georeferenced cultural heritage location map

Local Scale: Zaragoza City

To evaluate whether the implementation of collaborative cultural heritage mapping in the classroom facilitates the acquisition of digital geographic competencies and fosters a more critical and sustainable citizenship in relation to its environment—specifically in the city of Zaragoza—three structured educational experiences were designed (Fig. 3). These experiences were developed to integrate both pedagogical and methodological approaches that align with competency-based learning and geographic literacy.

The design of the educational interventions considered multiple dimensions:

(1) Educational levels: The study encompassed different stages of formal education to evaluate the impact of collaborative mapping across various cognitive and formative levels. The selected groups included: (1) 1st-year lower secondary education (K7): early-stage learners developing foundational geographic skills; (2) 3rd-year lower secondary education (K9): intermediate learners with a broader conceptual understanding of spatial relationships and urban dynamics; (3) Future secondary school teachers (Master's in Geography and Histiory Teacher Education): advanced learners engaging with pedagogical strategies for implementing digital geographic tools in educational settings.

(2) Heritage typologies: The mapping activities were structured around distinct types of cultural heritage to analyze how different heritage contexts influence geographic learning and critical thinking: (1) Contemporary heritage: The legacy of the International Exhibition *Expo Zaragoza 2008*, representing contemporary urban transformations and their socio-environmental impact with iconic architecture integrated in the river banks and a new suburban park; (2) Consolidated natural heritage inside buil environment: José Antonio Labordeta Park, created in the third decade of the twentieth century, as an example of green spaces within urban environments, fostering discussions on sustainability, public space, and urban ecology; (3) Consolidated cultural heritage: Zaragoza's Historic Center, a case study in urban history, landmark conservation, and cultural identity of the middle age and early modern period buildings (churches and convents, palaces…) and urban scene.

For the analysis of teaching praxis, a mixed-methods approach was used, combining quantitative evaluation rubrics with a qualitative analysis of student responses. The rubrics (based on a scale of 1–5), aligned with competency-based learning models, allowed for an objective measurement of students' acquisition of digital geographic skills and critical citizenship competencies. Statistical techniques were applied to assess trends in learning outcomes, while teachers' prior evaluations provided additional validation. The qualitative analysis examined students' reflections to assess their critical thinking, spatial citizenship awareness, and understanding of sustainable development (De Miguel 2021, 2025). This approach provided insights into students' engagement with collaborative mapping, the challenges they faced, and their perspectives on urban sustainability and heritage conservation.

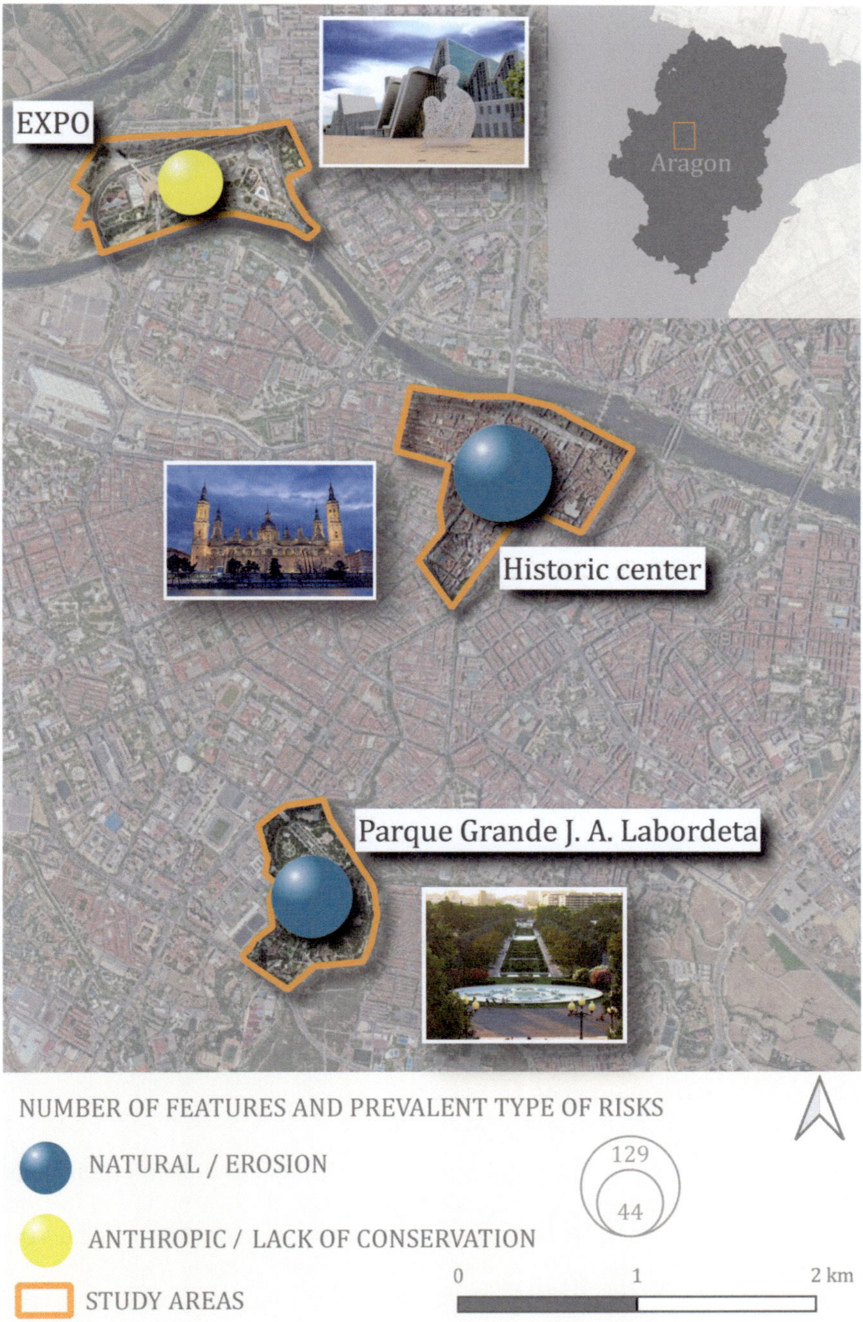

Fig. 3 Map of the educational field trips in Zaragoza

In all cases, an activity was designed based on educational itineraries, following a structured approach that encouraged both guided and autonomous learning (Fuller et al. 2006; Souto 2018; Morote Seguido 2019). The structure consisted of three key phases:

(1) Pre-fieldwork sessions: Before the on-site activity, students attended introductory sessions covering the necessary theoretical and methodological framework. These sessions explained the objectives of the activity, the significance of place and space, and the use of digital mapping tools.
(2) Semi-guided mapping during the fieldwork: The on-site activity was designed as a pseudo-guided mapping experience, balancing structured exploration with autonomous student participation. While facilitators provided initial guidance, students were encouraged to engage actively, document observations, and critically assess the heritage sites' conditions, fostering independent inquiry and spatial awareness.
(3) Post-fieldwork data processing and analysis: After the fieldwork, students engaged in data analysis and reflection, evaluating the condition of the mapped heritage sites. This phase included discussions on heritage conservation strategies and revalorization proposals, where students formulated solutions for sustainable heritage management based on their field observations (Table 1).

Results

This section presents the results obtained, structured around the two main objectives that guide the study, each analyzed at two different scales. First, the typology and valuation of heritage elements by users of ArcGIS Survey123 are examined, combining quantitative and qualitative approaches. Second, the participants' outcomes after the educational intervention are explored, focusing on the acquisition of spatial competencies, awareness of cultural heritage, and the promotion of sustainable spatial citizenship.

Mapping Spain's Cultural Heritage

A quantitative classification of 1098 heritage elements reveals a significant predominance of tangible heritage (Table 2), which constitutes 91.98% (1010) of the total dataset. In contrast, intangible heritage accounts for only 8.02% (88), indicating that heritage classification is largely oriented towards material aspects. This distribution suggests a potential bias in heritage recognition, where physical structures and artifacts receive greater emphasis compared to cultural practices, traditions, and oral expressions.

Regarding the perceived importance of heritage elements (Table 3), the majority were classified as either "important" (35.52%) or "very important" (32.60%), while

Table 1 Summary table of the three collaborative mapping educational interventions

Expo Zaragoza 2008 Number of students: 18 Average age of students: 14 years Number of georeferenced heritage assets: 44	
José Antonio Labordeta Park Number of students: 20 Average age of students: 15 years Number of georeferenced heritage assets: 87	
Historic Center Number of students: 60 Average age of students: 25 years Number of georeferenced heritage assets: 129	

Table 2 Georeferenced tangible and intangible heritage in the project

	N	%
Tangible heritage	1010	91.98
Intangible heritage	88	8.02
Total	**1098**	**100**

an additional 22.59% were considered "moderately important." In contrast, 9.29% of participants selected heritage elements that they personally regarded as "of little importance" or "not important at all."

Table 3 Degree of importance of georeferenced heritage

	N	%
Very important	358	32.60
Important	390	35.52
Moderately important	248	22.59
Slightly important	90	8.20
Not important	12	1.09
Total	**1098**	**100**

Table 4 Analysis of the open-ended response: why is this heritage asset important?

	N
Others	9146
Symbolic/social value	65
Historical value	56
Cultural value	44
Natural value	31
Identity value	21
Economic	13
Total	**9376**

From a scientific perspective, the classification was conducted through quantitative content analysis, systematically grouping terms based on predefined heritage value categories. The dataset contained a total of 9376 occurrences, which were assigned to their respective categories using keyword matching and semantic association techniques. The frequency distribution was then analyzed to determine the relative importance of each category (Table 4).

The analysis of heritage value categorization reveals that the majority of terms fall under the "Others" category, indicating a diverse range of concepts not specifically tied to a particular value classification. Among the defined categories, "Symbolic/Social Value" holds the highest frequency, highlighting the significance of heritage as a representation of collective meaning and identity. "Historical Value" follows closely, emphasizing the importance of historical elements in patrimonial recognition. "Cultural Value" and "Natural Value" also play notable roles, reflecting the relevance of cultural heritage and environmental aspects. Lastly, while less frequent, the "Identity Value" category reinforces heritage's role in fostering a sense of belonging and connection to traditions. This classification provides insight into how different heritage dimensions contribute to its broader societal significance.

The quantitative analysis of risk natural factors affecting heritage sites (Table 5), reveals that erosion (36.16%) and landslides (30.87%) are the predominant threats. Chemical action (11.84%) and seismic risk (3.73%) represent secondary hazards, while other factors (17.40%) contribute to additional degradation. This distribution may be influenced by a bias in participants' perception or even by a lack of clarity in

Table 5 Natural factors affecting heritage sites

	N	%
Erosion	397	36.16
Landslide	339	30.87
Chemical action	130	11.84
Seismic risk	41	3.73
Other	191	17.40
Total	**1098**	**100**

the survey design, which could affect the accurate identification of risk factors. Such limitations highlight the importance of refining assessment methodologies to ensure a more objective and comprehensive evaluation of heritage threats.

The quantitative analysis of anthropic risk factors affecting heritage preservation (Table 6) indicates that lack of conservation (44.26%) is the primary threat, followed by vandalism (21.95%). Inappropriate use (11.66%) and the threat of demolition or alteration due to urban pressure (10.20%) also pose significant risks. Additionally, other factors (11.93%) contribute to the degradation of heritage assets.

The quantitative and qualitative assessment of heritage elements underscores the predominance of tangible heritage in classification processes, revealing potential biases in recognition and conservation efforts. The varying degrees of perceived importance highlight the diverse ways in which communities value heritage, emphasizing the need for inclusive preservation strategies. Moreover, the identification of both natural and anthropic threats—particularly erosion, lack of conservation, and urban pressure—demonstrates the urgency of implementing targeted management policies. These findings call for a more comprehensive and interdisciplinary approach that integrates cultural, historical, environmental, and social dimensions to ensure the sustainable protection of heritage assets.

Table 6 Atrophic factors affecting heritage sites

	N	%
Lack of conservation	486	44.26
Vandalism	241	21.95
Inappropriate use	128	11.66
Threat of demolition or alteration due to urban pressure	112	10.20
Other	131	11.93
Total	**1098**	**100**

Towards a Sustainable Citizenship: The Case Study of Zaragoza

This section presents the results obtained regarding the acquisition of: (i) spatial and geographic competencies; (ii) awareness of cultural heritage; and (iii) sustainable spatial citizenship. A pretest-post-test design was used with natural groups, without the presence of a control group. The procedure was primarily based on three actions: first, an initial evaluation was conducted by the teaching staff to determine the baseline level; second, an educational intervention was carried out using collaborative mapping; and finally, after the completion of the courses, a post-test evaluation was also conducted by the teaching staff to assess the students' progress and final status.

The study results show that each intervention had a differential impact on the competencies evaluated, depending on the context in which it was conducted.

1. Expo Zaragoza 2008 (Table 7): the competency with the greatest variation was cultural heritage awareness, showing a significant increase in the post-test (2.8 $\rightarrow$ 4.6, $p = 0.0005$). This indicates that the setting of an expository event and access to informational materials facilitated an understanding of the heritage value of modern urban spaces. A notable improvement was also observed in spatial-geographic skills (3.2 $\rightarrow$ 4.5, $p = 0.001$), suggesting that collaborative mapping helped strengthen orientation and environmental analysis skills.
2. *Parque Grande José Antonio Labordeta* Park (Table 8): the competency that improved the most was sustainable spatial citizenship, with a remarkable increase (3.3 $\rightarrow$ 4.7, $p = 0.0001$), indicating that the relationship with the park's natural and cultural environment allowed students to reflect on sustainability and public space management. Cultural heritage awareness also showed a significant improvement (2.9 $\rightarrow$ 4.5, $p = 0.001$), suggesting that the combination of historical and natural elements in the park helped reinforce the perception of heritage value.
3. Historic Center (Table 9): the competency with the greatest variation was spatial-geographic skills, with a significant increase (3.0 $\rightarrow$ 4.3, $p = 0.001$). This suggests that the complexity of the urban layout and the rich heritage of the historic center favored the development of spatial skills. A significant growth was also observed in cultural heritage awareness (3.0 $\rightarrow$ 4.4, $p = 0.002$), indicating that

Table 7 Adjusted summary table of interventions and competencies

Expo Zaragoza 2008	Pre-test mean score	Post-test mean score	Significance level (p-value)
Spatial-geographic skills	3.2	4.5	0.001
Cultural heritage awareness	2.8	4.6	0.0005
Sustainable spatial citizenship	3.3	4.7	0.0001

Table 8 Adjusted summary table of interventions and competencies

Parque Grande José Antonio Labordeta Park	Pre-test mean score	Post-test mean score	Significance level (p-value)
Spatial-geographic skills	3.1	4.4	0.001
Cultural heritage awareness	2.9	4.5	0.002
Sustainable spatial citizenship	3.2	4.6	0.0005

immersion in a dense heritage context facilitated the appreciation of the historical and architectural value of the environment.

In addition to these competencies, detailed descriptive analyses have been conducted on the previous scale. However, we would also like to present the map showing the types of risks detected in each study area during the Zaragoza experiences $N = 260$ (Fig. 4). The map analysis reveals that natural and anthropic risks are simultaneously represented across all areas, suggesting a complex interaction between environmental and human factors. There are 33 features where natural risks outweigh anthropic ones, whereas in 32 cases, anthropic risks are more significant. Finally, 195 items present an equal distribution of natural and anthropic risks. This combination is not random but rather a result of territorial, urban, and environmental dynamics that amplify the vulnerability of heritage assets and urban spaces. Furthermore, from an educational perspective, this may contribute to the acquisition of spatial citizenship by fostering an understanding of the intricate relationship between environmental processes and human actions.

Last, it is important to highlight that the teachers involved in the Expo 2008 and *Parque Grande J.A. Labordeta* Park experiences emphasized the high level of student engagement during the session following the fieldwork. Secondary education students were able to develop their own critical thinking regarding the state of conservation of the mapped heritage assets and, at the same time, felt empowered to propose solutions. For example, they suggested the reuse of public space within the International Expo site or the creation of tourist itineraries through the cultural heritage assets of the park.

Table 9 Adjusted summary table of interventions and competencies

Historic Center	Pre-test mean score	Post-test mean score	Significance level (p-value)
Spatial-geographic skills	3.0	4.3	0.001
Cultural heritage awareness	3.0	4.4	0.002
Sustainable spatial citizenship	3.1	4.5	0.001

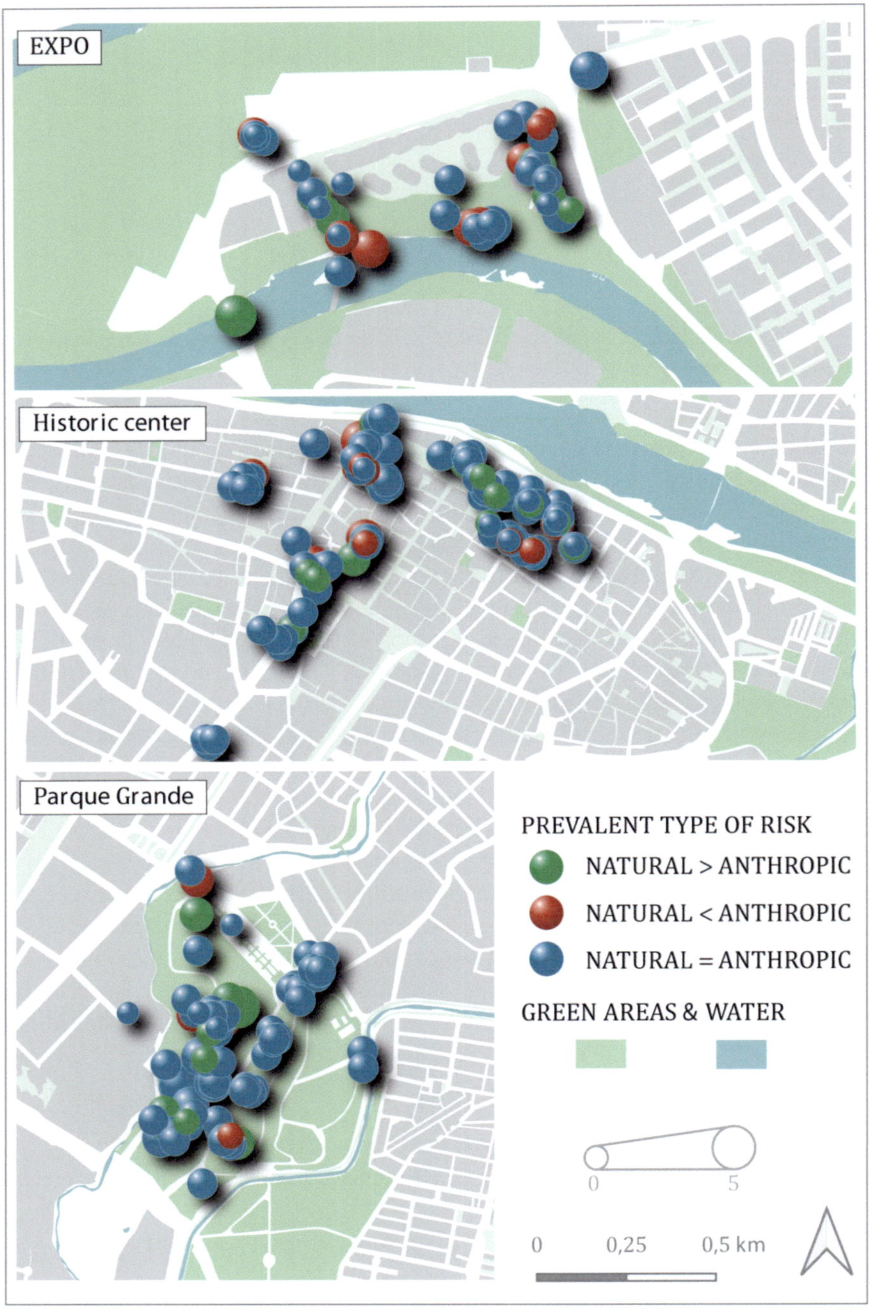

Fig. 4 Map of the educational field trip in Zaragoza

Discussion and Conclusions

This study reaffirms that Geography, History, and Heritage are not only taught in the classroom but are better. Their integrated and interdisciplinary nature enhances comprehension and engagement. Specifically, the main conclusions drawn from this research are: This study reaffirms that Geography, History, and Heritage are not only taught in the classroom but are better understood and internalized through experiential learning beyond it. Their integrated and interdisciplinary nature as school subjects enhances deeper comprehension and engagement. Specifically, the main conclusions drawn from this research are.

(a) Collaborative mapping fosters spatial citizenship and civic engagement. The use of geospatial tools in heritage education facilitates a deeper understanding of cultural landscapes and empowers citizens to actively participate in heritage conservation. Students and local communities develop a sense of spatial responsibility, recognizing the significance of preserving historical and cultural assets.

(b) There is a bias towards tangible heritage, highlighting the need for a broader recognition of heritage. The predominance of tangible over intangible heritage (91.98 vs. 8.02%) in the mapping process suggests that built heritage receives greater attention than oral traditions, folklore, and cultural practices. This highlights the need for more inclusive approaches to ensure a balanced representation of all forms of heritage.

(c) Educational interventions strengthen digital geographic competencies. The CIVITAS III project and related initiatives demonstrate that integrating GIS-based collaborative mapping into educational curricula enhances students' spatial literacy, critical thinking, and problem-solving skills. This interdisciplinary approach aligns with competency-based learning models and promotes sustainable citizenship practices.

(d) Heritage conservation should be embedded in sustainable urban planning practices. Cultural heritage goes beyond being a historical object or mere memory; it is an essential part of everyday life in modern society. The findings emphasize the need to incorporate heritage management into territorial development policies, ensuring that cities balance modernization with historical preservation to promote inclusive, resilient, and culturally enriched urban environments.

Incorporating cultural heritage into territorial development is not merely about preserving the past but also shaping the future. By combining historical and geographical knowledge with modern tools such as geospatial technology, societies can foster more inclusive, informed, and culturally enriched urban and rural environments. In this way, cultural heritage becomes not just a relic of history but a dynamic and integral element of contemporary life and sustainable development.

This chapter highlights the crucial role of collaborative mapping in the preservation of endangered cultural heritage and its integration into sustainable urban development. By engaging students and local communities in geospatial documentation and analysis, the research demonstrates how digital tools can enhance spatial awareness, foster civic participation, and promote informed decision-making for heritage conservation. The findings reinforce the idea that Geography, History, and Heritage are best understood through geo-inquiry learning, bridging the gap between theoretical knowledge and practical application. Additionally, the categorization of heritage values underscores the diverse ways in which cultural assets contribute to collective identity, historical continuity, and environmental awareness. Ultimately, this study reaffirms that cultural heritage is not just a legacy of the past but a key driver of sustainable and inclusive urban and territorial planning for the future.

Acknowledgements This research was funded by the project *Digital competences, learning processes and cultural heritage awareness: quality education for sustainable cities and communities* (PID2020- 115288RB-I00), financed by the Spanish State Research Agency and the Spanish Ministry of Science and Innovation and by *ARGOS-IUCA research group*, funded by Aragon's Government with code (S23_50R). Additionally, Juan Mar-Beguería is supported by a predoctoral grant financed by the University of Zaragoza (code PI-PRD-001/2023).

References

Álvarez-Otero J, De Lázaro-Torres ML (2018) Education in sustainable development goals using the spatial data infrastructures and the TPACK model. Educ Sci 8(4):171. https://doi.org/10.3390/educsci8040171

Amichai-Hamburger Y (2008) Potential and promise of online volunteering. Comput Hum Behav 24(2):544–562

Bokova IG (2023) UNESCO moving forward the 2030 agenda for sustainable development. UNESCO. Available online https://unesdoc.unesco.org/ark:/48223/pf0000247785

Buzo-Sánchez IJ, Mínguez C, De Lázaro-Torres ML (2022) Expert perspectives on GIS use in Spanish geographic education. Int J Digit Earth 15(1):1205–1219. https://doi.org/10.1080/17538947.2022.209613

Buzo-Sánchez IJ, Mínguez C, De Lázaro-Torres ML (2023) The potential of the SMART learning framework to design and implement geospatial curricula in the secondary classroom. J Geogr 122(6):141–154. https://doi.org/10.1080/00221341.2023.2261108

De Miguel R (2021) Geographical and geospatial competences from school education to higher education: The contribution of EUROGEO in the international projects. J Res Didactics Geogr 2(10):57–65

De Miguel González R (2024) Powerful geography and the future of geographic education. Dialogues Hum Geogr 14(1):5–8. https://doi.org/10.1177/20438206241229219

De Miguel R (2025) The discipline of geography and geography education. In Mitchell J, Bednarz (eds) Handbook for geography education. Springer. https://link.springer.com/book/9783031723650

De Miguel R, Donert K, Kostis K (eds) (2019) Geospatial technologies in geography education. Springer, Cham. https://doi.org/10.1007/978-3-030-17783-6

De Miguel R, De Lázaro ML (2020) WebGIS implementation and effectiveness in secondary education using the digital atlas for schools. J Geogr 119:74–85. https://doi.org/10.1080/00221341.2020.1726991

De Miguel R, Sebastián M (2022) Education on sustainable development goals: geographical perspectives for gender equality in sustainable cities and communities. Sustainability 14(7):4042. https://doi.org/10.3390/su14074042

Fuller I, Edmondson S, France D, Higgitt D, Ratinen I (2006) International perspectives on the effectiveness of geography fieldwork for learning. J Geogr High Educ 30(1):89–101
Giovene di Girasole E, Daldanise G, Clemente M (2019) Strategic collaborative process for cultural heritage. In: Calabrò F, Della Spina L, Bevilacqua C (eds) New metropolitan perspectives. ISHT 2018. Smart innovation, systems and technologies, vol 101. Springer, Cham. https://doi.org/10.1007/978-3-319-92102-0_38
Martínez de Pisón E (2010) Saber ver el paisaje. Estudios geográficos 71(269):395–414
Morote Seguido ÁF (2019) Las salidas de campo en España como recurso didáctico para la enseñanza de la Geografía. Una revisión bibliográfica. Geographicalia 71:27–49. https://doi.org/10.26754/ojs_geoph/geoph.2019714142
Sebastián López M, De Miguel González R (2017) Educación geográfica 2020. Iberpix y Collector for ArcGis como recursos didácticos para el aprendizaje del espacio. Didáctica Geográfica 18(313–316)
Sebastián-López M, De Miguel González R (2020) Mobile learning for sustainable development and environmental teacher education. Sustainability 12(22):9757. https://doi.org/10.3390/su12229757
Sebastián López M, Kratochvíl O, De Miguel González R (2023) Geographic education and spatial citizenship: collaborative mapping for learning the local environment in a global context. In: Re-visioning geography, pp 177–191
Serrano A, Sebastián M (2024) Estrategias educativas para la lucha contra la despoblación: una propuesta transversal a través de las asignaturas de geografía e historia en institutos rurales de Aragón. In: García-Delgado FJ, Hernández-Hernández M, Martínez-Puche A, Hortelano Mínguez LA, Zapata Hernández VM (eds) (Con)textos de desarrollo local. Dos décadas de Geografía aplicada (2003–2023). Asociación Española de Geografía (AGE)-Grupo de Trabajo de Desarrollo Local de la Asociación Española de Geografía (GTDL-AGE). https://doi.org/10.21138/pgP.2024.lc
Sharpe B, Bahbahani K, Tu Huynh N (2016) Teaching geographical thinking (revised and expanded edition). In: The critical thinking consortium
Stamenov S, Stamenova V (2021) Geoinformation technologies for conservation of cultural heritage. Geoinformation technologies for conservation of cultural heritage. In Ioannides M, Fink E, Cantoni L, Champion E (eds) EuroMed 2020: Digital heritage. Progress in cultural heritage: documentation, preservation, and protection. Lecture notes in computer science. Springer, Cham, p 12642
Souto XM (2018) School geography: institutional aspirations and classroom experiences. *Boletín* de *la Asociación* de Geógrafos Españoles
UNESCO (2021) Preparing world heritage nomination, 2nd edn. UNESCO
UNESCO (2023) Convention for the safeguarding of the intangible cultural heritage. UNESCO
Van Boxtel C, Grever M, Klein S (2016) Sensitive pasts. Questioning heritage in education. Berghahn Books

Cross-Border Cultural and Economic Cooperation: A Study of Gornja Radgona in Slovenia and Bad Radkersburg (Radgona) in Austria

Peter Kumer and Uroš Horvat

Abstract This chapter explores the impact of cross-border cultural and economic cooperation between the towns of Bad Radkersburg (Radgona) in Austria and Gornja Radgona in Slovenia. The study involved geography students conducting fieldwork, where they interviewed local bystanders using the ESRI mobile app, Survey123, to gather data on various aspects of cooperation between these border towns. This study focuses on how such cooperation contributes to intercultural learning, enhances knowledge of languages, and preserves cultural heritage. The primary objective is to understand the dynamics of cross-border interactions in the realms of culture, economy, sport, and leisure, and to examine how these interactions promote intercultural learning and bilingualism. The Austrian part of the Bad Radkersburg municipality (Radgona) is home to a Slovenian minority that has historically been suppressed. This chapter investigates how new cross-border interactions contribute to the visibility and recognition of this minority group. The data collected by students reveal insights into the everyday experiences and challenges faced by the local population, emphasizing the role of geospatial technologies in heritage education. This chapter discusses the methodology, including the use of Survey123 for data collection, and analyzes the findings to highlight the importance of heritage in fostering a sense of community and identity. It also illustrates how geospatial technologies can be effectively integrated into educational practices to enhancestudents' understanding of heritage and its diverse elements. By showcasing the positive outcomes of cross-border cooperation, this study aims to contribute to the broader discourse on heritage education and the application of geospatial technologies in understanding and preserving cultural heritage.

P. Kumer (✉) · U. Horvat
University of Maribor, Faculty of Arts, Department of Geography, Maribor, Slovenia
e-mail: peter.kumer@um.si

U. Horvat
e-mail: uros.horvat@um.si

P. Kumer
University of Ljubljana, Faculty of Social Sciences, Research Center for Spatial Sociology, Ljubljana, Slovenia

R. de Miguel González and P. Rivero Gracia (eds.), *Geospatial Technologies for Heritage Education*, Key Challenges in Geography,
https://doi.org/10.1007/978-3-031-96372-8_7

Keywords Cross-border cooperation · Cultural heritage · Slovenian minority · Bilingualism · Interregional cooperation · Geospatial technologies

Introduction

Cross-border cooperation is one of the key elements of the European Union's policy. By supporting sustainable development along borders, it aims to reduce disparities in living standards and address common challenges across these borders (Gerfert 2009). Significant economic and social differences between regions and countries hinder the development of the Union as a whole. Cross-border cooperation, within the framework of European Territorial Cooperation, is part of the common European cohesion policy, which supports the balanced development of the entire European Union. The fundamental objectives of cross-border cooperation programs are to promote economic and social development in border areas, address common challenges in the fields of the environment, public health, safety, and security, and establish better conditions for the mobility of people, goods, and capital.

National minorities can play an especially important role in the development of border regions, as they can serve as a connecting link between neighboring nations. Zupančič (2005) notes that minorities were, in the past, primarily due to the desires of national ideologies for the ethnic unification of state territories, initially subjects of political and strategic bargaining or even interstate conflicts, until they eventually emerged nowadays as an important connecting element in interstate and interregional relations during the European integration processes.

The process of European integration and the enlargement of the European Union have simplified and therefore increased cross-border communication and interregional cooperation. With the open European labor market and the free movement of labor, international mobility has increased, and a more unified communication space has also been formed. Minority populations have proven to be important promoters and carriers of various activities. In some cases, minority members, their organizations, and institutions have played an irreplaceable role as so-called "bridges" in connecting regions, countries, and nations (Zupančič 2005). Additionally, Klotz (2019) notes that cross-border cooperation is of fundamental importance in border areas where indigenous or national minorities live, as they often have a special interest in establishing contacts with individuals or authorities in neighboring countries due to cultural, linguistic, and historical proximity. This is particularly important in the European context for members of the Slovenian minority, as they live in all neighboring countries. Unlike other Eastern European socialist states, the former Yugoslavia (and thus Slovenia) had open borders with Western countries as early as the mid-1960s, and in the later period, there were already some cases of successful cross-border cooperation.

The aim of this chapter is to demonstrate the impact of cross-border cultural and economic cooperation using the example of two small towns along the Slovenian Austrian border in Styria (Steiermark/Štajerska): Bad Radkersburg (Radgona) in

Austria and Gornja Radgona in Slovenia, separated, but also connected by the Mura River. Geography students from the University of Maribor (Slovenia) were involved in data collection, conducting fieldwork, and using the ESRI Survey123 mobile application to interview local passersby to gather data on various aspects of cooperation between the border towns. The research focused primarily on determining how cross-border cooperation contributes to intercultural understanding, strengthens language skills, and preserves cultural heritage. The central aim of the research is to understand the dynamics of cross-border interactions in the fields of culture, economy, sports, and leisure, and to examine how these interactions promote intercultural understanding and bilingualism. It is also important to note that a small Slovenian national community lives in Bad Radkersburg (Radgona) and its surroundings, a group that has historically been oppressed and unrecognized, only beginning to actively advocate for the rights of the Slovenian population in Austrian Styria a few decades ago, including promoting Slovenian language education and active cross-border cooperation in culture and science. Therefore, one of the objectives is also to determine the role of the minority in new cross-border interactions and population mobility on both sides of the border, and whether these contribute to the visibility and recognition of the minority community.

With the collected data, we aimed to gain insight into the everyday experiences and challenges faced by the local population, with a focus on the role of geospatial technologies in educating about the importance of heritage (in our case, the use of the autochthonous Slovenian language) and in fostering a sense of community and (territorial) identity. By showcasing the positive outcomes of cross-border cooperation, we hope to contribute to the broader discourse on heritage education and the use of geospatial technologies in understanding and preserving cultural heritage.

Historical Background: Formation of the Slovenian State Border

The ethnic structure of the population in the present-day territory of Slovenia and its surroundings is the result of centuries of historical processes, shaped by wars, economic and spatial development, numerous political changes, and migrations. These processes have formed the landscape of Slovenian culture, defined its borders, and outlined the fundamental features of its ethnic composition. After World War I, when states for individual nations began to form on the territory of the former Austro-Hungarian Empire, the borders between them could not be drawn strictly based on ethnic or national principles due to the intermingling of ethnically mixed areas with regions of homogeneous settlement by individual nations. As a result, through three phases of national state formation (after World War I, after World War II, and after the breakup of the former Yugoslavia in the 1990s), borders were established that still run through ethnically mixed areas today (Klemenčič 1994).

Slovenia, in the process of forming national state is considered a young Central European country, as the Slovenian nation gained its statehood for the first time in history only three decades ago. The independent Slovenian state was established on June 26, 1991, emerging from the Republic of Slovenia, which was then one of the republics within the former Socialist Federal Republic of Yugoslavia. The independent state is the result of the efforts of several generations of Slovenians in this part of Europe. For several centuries, most of the Slovenian national territory was part of the Austrian Empire, which was transformed into the Austro-Hungarian Empire in 1867. Hungarian statehood also triggered a wave of Pan-Slavism among the Slavs for the first time in the second half of the nineteenth century (Prunk 2006). After World War I, most of Slovenia's territory became part of the Kingdom of Serbs, Croats and Slovenes (Kingdom of SHS), which was in 1929 renamed into the Kingdom of Yugoslavia. After World War II, the Slovenian nation made significant economic and cultural progress within the frame of the socialist federative Yugoslavia. However, the severe crisis of the Yugoslav communist system and the increased national tensions that culminated in the breakup of Yugoslavia in 1991 further motivated Slovenia to declare its independent statehood.

Slovenia is a relatively small country, covering an area of only 20,273 square kilometers and having nowadays a population of around 2.1 million people. Positioned in the southern part of Central Europe, it is truly a "border state," as more than half of its territory lies within a 25-km border zone (Zupančič 2017). The total length of its borders is 1334 km, of which approximately 50% borders Croatia, 25% Austria, 17% Italy, and 8% Hungary. Another way to measure Slovenia's border character is by calculating the kilometers of international border per 100 km^2 of state territory. In this regard, Slovenia ranks as the second country in Europe, with 5.7 km of border per 100 km^2 (Bufon 2004).

The Slovenian national territory is larger than the territory of the state itself. Members of the Slovenian national community live in the territories of all neighboring countries, which is a consequence of historical events and numerous political decisions made in the past, mostly by members of other nations. The border with present-day Austria was defined after World War I. Slovenians sought to include all Slovenian-populated territories within the Kingdom of SHS, but the Treaty of Saint-Germain in 1919 established only part of the state border between the two countries. For the area in the region of Carinthia (Kärnten/Koroška) where there was a significant ethnically mixed population, a plebiscite was held. In the 1920, approximately 59% of the population (due to complex political and economic reasons) decided to remain in Austria, and that lead to the establishment of the border along the Karavanke mountain range (Svenšek et al. 2020).

After World War I, the border with present-day Hungary was also established. The territory on the left bank of the Mura River (Prekmurje region), predominantly populated by Slovenians, had belonged to Hungary since the Middle Ages. Following 1918, a series of events led to the Treaty of Trianon in 1920 between the Kingdom of SHS and Hungary. Treaty reunited the Prekmurje region with the Slovenian national territory after nearly a thousand years.

The border with Italy was established only in the second half of the twentieth century. Following World War I, the Treaty of Rapallo in 1920 granted a large part of the Slovenian national territory in the western part of present-day Slovenia (which included more than a quarter of the Slovenian population at the time) to Fascist Italy. At the end of World War II, this territory was liberated by the Yugoslav army. However, the border was not formally established until 1954, when Yugoslavia regained parts of the Slovenian ethnic territory, including access to the Adriatic Sea near Koper, from Italy. The final resolution of the border with Italy was achieved only with the Osimo Agreements in 1975 (Robežnik 2024).

After the breakup of Yugoslavia, Slovenia retained its external borders with Italy, Austria, and Hungary upon declaring its new state in 1991, and newly established a border with Croatia, which follows the former republic boundary. This border is largely based on centuries-old demarcations, and for the most part, it has not been problematic, except for a few sections, most notably the Bay of Piran (Kladnik and Pipan 2008), where, despite international arbitration, a mutually agreeable resolution between the two countries has not yet been reached.

Slovenian National Minority in Neighboring Countries

As a result of historical facts, members of the Slovenian nation live as a minority in all neighboring countries. They play a significant role in shaping the border regions. Moreover, they act as a vital link between the two neighboring nations. It is estimated that in 1910 approximately 240,000 Slovenians lived in border areas in neighboring countries, with around 110,000 in Austria and about 120,000 in Italy (Zupančič 2005). Due to a series of historical changes, ongoing social processes of assimilation or adaptation to the majority population, and structural weaknesses in these areas (such as low population density, demographic vulnerability, underdeveloped infrastructure, lack of central functions, and poorer access to services), the number of Slovenians living in these border regions has decreased across all areas over the past century. Estimates indicate that in 2000, around 170,000 members of the Slovenian nation lived in border areas, with about 100,000 in Italy (in the regions of Trieste, Gorizia, and Udine), up to 60,000 in Austria (in Carinthia and Styria), around 5000 in Hungary (in the Raba region), and a few thousand along the Slovenian-Croatian border (Stergar 2001). Their numbers continue to decline, and recent estimates suggest that there are now around 150,000 Slovenians in these areas.

For this reason, it has been a historical priority for the Slovenian nation and state to preserve the minority communities and to promote their economic, social, and cultural development. Article 5 of the Constitution of the Republic of Slovenia explicitly addresses the protection of the autochthonous Slovenian national minorities in neighboring countries. At the same time, Slovenia is also committed in the same article to safeguarding the rights of the autochthonous Italian and Hungarian national communities within Slovenia. According to the results of the 2002 census, which was the last to determine the ethnic composition of the population, around

3,700 people in Slovenia identified Italian as their mother tongue, and about 7,700 identified Hungarian. Both national communities are guaranteed all rights under the Constitution of the Republic of Slovenia, regardless of the number of community members. Additionally, each community is granted one parliamentary seat in the National Assembly of the Republic of Slovenia, which is a rather unique solution for minority rights globally. Slovenia is therefore considered a country with some of the highest minority standards, both normatively and in practice.

On the other hand, especially in the past, the national life of Slovenians who remained in neighboring countries was extremely difficult, as these countries often failed to honor the promises of protection and the free development of the Slovenian minority that they made when signing peace or state treaties. Since this article deals with the area along the Slovenian-Austrian border, only the position of the Slovenian minority in Austria is presented below.

Austria committed itself under international law to guarantee the rights of the Slovenian minority in Carinthia and Styria through Article 7 of the Austrian State Treaty in 1955. However, this commitment has not been fully realized to this day, particularly in terms of ensuring visible bilingualism and an appropriate status for the Slovenian language in bilingual areas (Gov.si 2024). Today, in southern Carinthia, where bilingual education is in place, there are several bilingual primary schools and kindergartens, and over 100 Slovenian associations and organizations operate in the fields of culture, sports, youth, economy, agriculture, and other areas. Media and publishing houses that regularly produce content in the Slovenian language also play an important role in preserving the Slovenian language and identity (Gov.si 2024a). The majority of Slovenians in Carinthia live in the southern and eastern parts of the region. The Slovenians in Styria are less well-known, but because they form a significantly smaller and more assimilated community, the realization of their rights is much more challenging and slower.

Case Study Area: Radgonski Kot Region

When the border between Austria and the Kingdom of SHS was established, some territories populated by Slovenian people remained on the Austrian side, within the Austrian federal state of Styria (Steiermark/Štajerska). However, this area was significantly smaller in both size and the number of Slovenian inhabitants compared to the neighboring federal state of Carinthia (Kärnten/Koroška). From west to east, these areas included the former municipality of Sobota (Soboth), particularly in the settlements of Mlake (Laaken), Radvanje (Rothwein), Gradišče (Schlossberg), and Klanci (Glanz). Additionally, there were five densely populated settlements in the so-called Radgonski kot region (Klemenčič 1978), covering approximately 20 km^2 between the Mura River to the south, the Kučnica stream to the east, and the foothills towards Klek/Klöch to the north.

The regional center is the town of Radgona (today's Bad Radkersburg), where in 1910, the Slovenian population represented only about 2% of the population.

However, in the villages of the Radgonski kot area: Slovenska Gorica (Goritz), Zenkovci (Zelting), Dedonci (Dedenitz), Žetinci (Sicheldorf), and Potrna (Laafeld), a significant portion of the population (about 62%) identified Slovene as their spoken language (Rihtarič 2026). Besides the indigenous population, Slovenian immigrants from the nearby Slovenian region of Prekmurje also played an important role in this area (Rihtarič 2026).

According to the 1910 census, which recorded the number of inhabitants based on their spoken language, there were 2644 individuals with Slovene as their spoken language living in the indigenous settlement area of Austrian Styria, and 5146 across the entire region of Austrian Styria. In subsequent censuses, the number of Slovene speakers decreased significantly, with only 246 individuals recorded in 1991 (Zupančič 1997).

Although the Treaty of Saint-Germain in 1919 guaranteed a certain level of minority rights for minorities in Austria, these rights were not recognized for Slovenians in Styria. In contrast to Carinthia, where some minority rights were at least partially granted, Slovenians in Styria were denied any such recognition. As a result, the Slovenian minority in Austrian Styria had no opportunity for organized activity, did not develop a cultural life, and for a long time was cut off from contact with

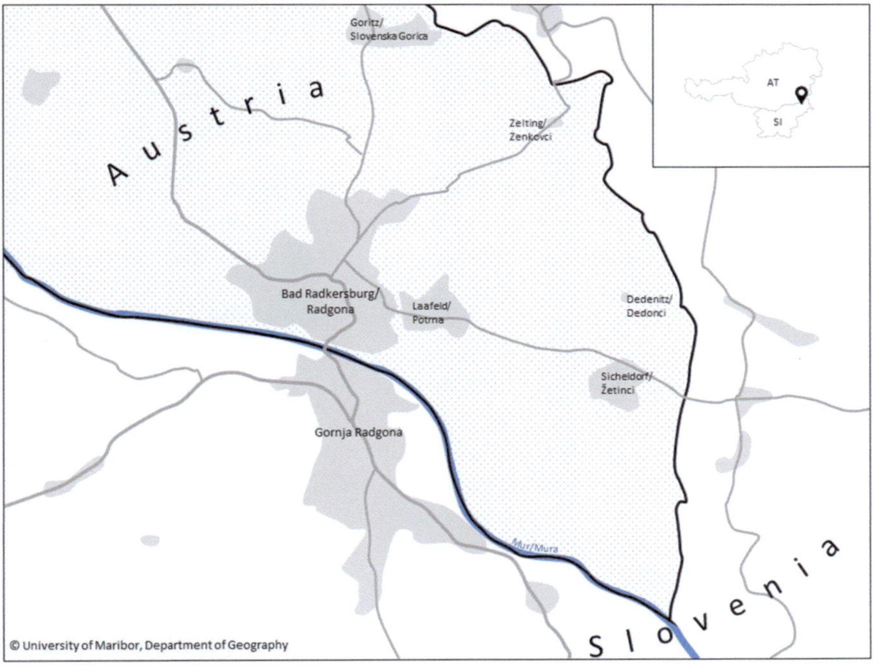

Fig. 1 Map of the case study area (Radgonski kot), showing Bad Radkersburg (Radgona) and Gornja Radgona as the locations where fieldwork was conducted, and highlighting five bilingual villages in the Radgonski kot in present-day Austria

their mother nation as well as from the Slovenians in Carinthia. They also lacked any connection to the Slovenian language (Jesih 1998).

Despite the rights enshrined in Article 7 of the Austrian State Treaty in 1955, the situation for Slovenians in Austrian Styria did not improve in the second half of the twentieth century. In fact, there was a threat that the minority might practically disappear due to assimilation. However, in 1988, some engaged individuals from both sides of the border succeeded in establishing the "Cultural Association Article 7 for the Styrian Slovenians". Despite facing significant challenges, the association survived and became a recognized organization for Styrian Slovenians. The association advocates for the rights of the Slovenian population in Austrian Styria, promotes Slovenian language education, and actively supports cross-border cooperation in the fields of culture and science. In 1995, the association acquired an old farmhouse in Laafeld (Potrna) near Bad Radkersburg (Radgona), renovated it, and in 1998 opened the cultural center Pavlova hiša (Pavelhaus). This center serves as a hub and cultural center for the monolingual and bilingual population in the area and is described as a place of dialogue by the association (Weitlaner 2023). Additionally, other Slovenian associations operate in Austrian Styria, particularly in Graz, including the Club of Slovenian Students and the Austrian Slovenian Friendship Society.

From the perspective of minority protection, cultural heritage, and especially the preservation of the Slovenian language in Austrian Styria, it is important to mention some efforts aimed at its scientific study. The greatest focus has been on Radgonski kot region, where researchers have primarily explored the dialectological characteristics of the language (Zorko 1989), issues of identity and language preference among Slovenian speakers (Križman 1987), and the linguistic aspects of the social and historical circumstances surrounding the development of the area (Haberl-Zemljič 2004). The most recent field research has aimed at confirming the existence and documenting the speech of this area. For a long time, it was believed that there were no longer any Slovenian speakers in this region, but these studies have reaffirmed their presence (Gostenčnik et al. 2022).

With the rapid economic development of Austrian Styria, the opening of borders, and the free movement of people and goods (especially following Slovenia's entry into the Schengen Area in 2007), more and more Slovenians are working and living in Austrian Styria. According to estimates by the Cultural Association Article 7 for Styrian Slovenians, around 6200 Slovenians now live in Austrian Styria, with approximately 3200 residing in Graz (Weitlaner 2023). This number does not include the indigenous Styrian Slovenians, who are estimated to number slightly over a thousand across the entire Styrian region.

The largest settlement in the border area of Austrian Styria is Bad Radkersburg (Radgona), located on the Mura River, which serves as both a border river and, in many places, the dividing line between the two language communities. The area has been inhabited since the twelfth century and was elevated to the status of an imperial fortress in the border region in the sixteenth century. In 1918, it was occupied by the Yugoslav army, but the territory north of the Mura River was awarded to Austria by the Treaty of Saint-Germain. The area also suffered significant damage during World War II. In 1969, a new bridge across the river reconnected both banks, coinciding with the significant development of the town of Gornja Radgona on the Slovenian side of the Mura River. In 1975, Bad Radkersburg was granted spa town status, and in recent years it has recorded around half a million tourist overnight stays annually (Bad Radkersburg 2024). Despite this, the town's population has been declining since the early twentieth century. In 1910, the town had around 2800 inhabitants, about 2000 in 1971, and only 1400 residents in 2024. The entire municipality of Bad Radkersburg currently has approximately 3200 inhabitants (Bad Radkersburg 2024).

On the Slovenian side of the border, the town of Gornja Radgona began to develop more rapidly after World War II, supported by its rich viticultural hinterland and economic development. Today, the town has about 3100 residents, while the entire municipality has around 8,500 inhabitants, surpassing its neighboring Austrian counterpart. Cooperation between the towns on both sides of the border has been ongoing for many years, primarily focusing on culture and sports, which has, in turn, boosted tourism. The two towns are twinned, and there is also lively economic cooperation, particularly in trade and services. The free movement of goods and services has accelerated this cooperation in the last decade, and it has also enabled many Slovenian residents to find employment on the Austrian side, with many commuting daily across the border for work.

Methods

Survey Design

The survey was designed for fieldwork, an essential research method in geography. It includes questions about the frequency and intention of border crossings, cross-border cooperation, language issues, and the social and demographic status of respondents. The questionnaire was specifically adapted for the ESRI ArcGIS Survey123 mobile application, with additional questions to capture the time and location of the interview, as well as photographs (Fig. 2).

The survey was tested by second-year undergraduate students of geography at the University of Maribor, Slovenia. Although not yet experts, they were trained in the Social & Cultural Geography course to conduct field surveys. They were also tasked with performing literature reviews and conducting in-person interviews to understand the status and historical and contemporary issues of the Slovene minority in Austria, as well as Slovene-Austrian cross-border cooperation. At the conclusion of this fieldwork, students were asked to complete a brief questionnaire reflecting on what they learned, what they enjoyed, and the challenges they encountered.

Fig. 2 A student conducting a fieldwork survey in Gornja Radgona with a passerby using a mobile app

Study Sample

This survey was conducted by students during fieldwork on 17 May 2024, in two intervals: 8:00–9:00 AM and 11:00 AM–12:00 PM. Using a location-aware app to record responses, the students were divided into four groups with an equal number of interviewers to ensure spatial distribution across both border towns: Bad Radkersburg (Austria) and Gornja Radgona (Slovenia) (Fig. 3). German-speaking students were allocated to the Austrian side, and questions were written in both languages for additional support.

The brief questionnaire took approximately 5 min to complete. In total, 92 respondents participated in the survey. To ensure a balanced representation of respondents from both towns, as well as equal representation of both genders and all age groups, the instructor (author of this chapter) monitored real-time survey results on the ESRI portal and used a WhatsApp community to inform students about underrepresented groups that needed to be targeted.

However, there are some biases in the sample as the survey was conducted during morning working hours on a Friday. The largest age group interviewed on the streets was 60 years or older, possibly due to the working-age population being at their jobs

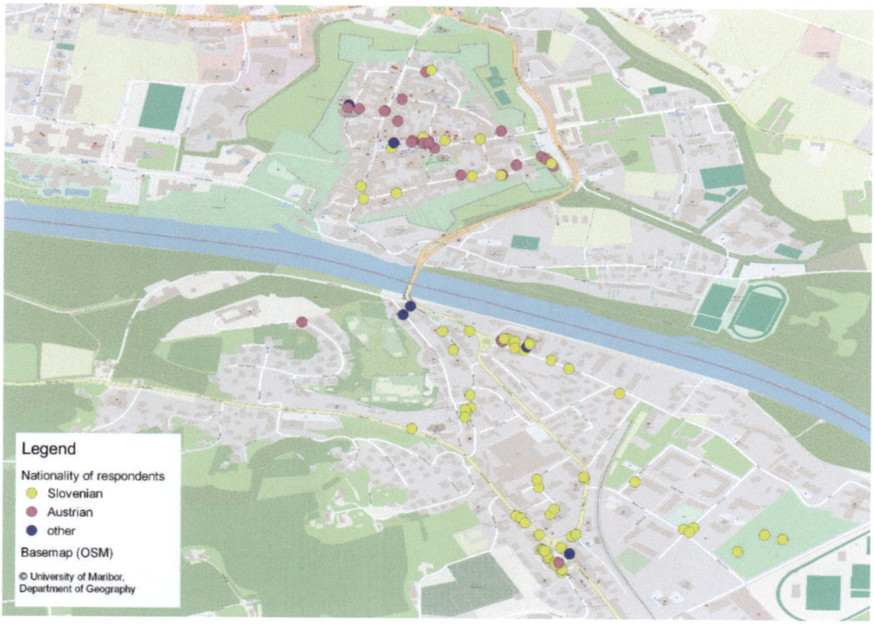

Fig. 3 Locations of respondents in Bad Radkersburg (Radgona) and Gornja Radgona, with colors representing their nationality (Slovenian, Austrian, and other)

(see Table 1). Additionally, while 86.2% of respondents on the Slovenian side were Slovenian, only 58.8% of respondents on the Austrian side were Austrian. This may be due to the fact that a large share of the Slovenes interviewed on the Austrian side were working migrants (although the presence of Slovene ethnic minority on Austrian side is also an important factor), while fewer Austrians, who typically visit the Slovenian side during weekends and after work hours, were present.

Measures

The survey collected data on various sociodemographic factors, including age, gender, nationality, and education level. We also collected data on border crossing frequency, reasons for crossing, language proficiency, etc. The app itself produced the following responses: geographical locations and times of interviews. These measures provided a comprehensive understanding of the respondents' backgrounds and were crucial in analyzing the patterns and trends in the responses.

Table 1 Descriptive statistics of research sample

Demographic variable			Frequency	%
Age groups	Under 18		3	3.3
	18–29 years old		8	8.7
	30–44 years old		19	20.6
	45–59 years old		24	26.1
	60 years and older		38	41.3
Gender	Males		42	45.7
	Females		50	54.3
Nationality	Bad Radkersburg, 34	Slovenians	12	35.3
		Austrians	20	58.8
		Others	2	5.9
	Gornja Radgona, 58	Slovenians	50	86.2
		Austrians	4	6.9
		Others	4	6.9
Education	Completed primary school		7	7.6
	Completed 3-year secondary school		14	15.2
	Completed 4-year secondary school		50	54.3
	Completed 1st cycle UNI		10	10.9
	Completed 2nd cycle UNI		8	8.7
	Other		3	3.3

Analyses

For the analyses, data were assessed through both descriptive statistics and multivariate regression models to understand the relationship between different variables. The study specifically focused on the frequency and intention of border crossings, cross-border cooperation, language issues, and the social and demographic status of respondents. The analyses aimed to identify significant patterns and correlations within the collected data, providing insights into the factors influencing cross-border interactions.

Table 2 Responses by Austrians and Slovenians regarding the types of activities they cooperate in across the border

	Austrians, frequency	Slovenians, frequency
Cultural and artistic activities	8	8
Shopping	17	38
Trade or business	5	7
Recreational and sports activities	5	12
Other activities	3	13

Results

Frequency and Reasons for Crossing the Border

To understand border crossing behavior, we compared border crossing between Slovenes and Austrians. The comparison of border crossing frequency between Slovenes and Austrians reveals distinct patterns. Slovenes are more likely to cross the border daily (31.6%) and monthly (24.6%) compared to Austrians, who show a higher tendency to cross the border weekly (41.7%). Austrians also exhibit a higher frequency of very rare crossings (12.5%) compared to Slovenes (1.8%). Conversely, Slovenes cross the border a few times a year (15.8%) more frequently than Austrians (8.3%). These differences highlight that Slovenes tend to have more frequent and consistent border crossings, while Austrians have a more varied crossing pattern with a notable proportion crossing weekly and not unsurprisingly—very rarely.

To understand whether this is related to the reasons for border crossings, we examined the main reasons provided in the open-ended question of the questionnaire. We identified four primary reasons for border crossings (shopping 42.4%, work 25%, recreation and sport 8.7%, visiting friends and relatives 7.6%, and transiting to other places 5.4%).

Among the pre-defined answers (Table 2), different types of cross-border cooperation are reported by each nationality. Interest in cross-border retail is similar between both groups. Cooperation in recreational and sports activities is also nearly equal among Austrians and Slovenians. However, Austrians report slightly more participation in trade or business interactions, as well as in cultural and artistic activities. On the other hand, Slovenians appear to engage more in other activities (particularly work or trips/travel).

Border Crossing by Location and Nationality

The analysis of border crossing reasons **by location** reveals that more people in Gornja Radgona cross the border for shopping (50%) compared to Bad Radkersburg

(Radgona), where only 41.2% of respondents cross for shopping. Regarding work, in Gornja Radgona 22.4% cross the border for work compared to Bad Radkersburg where we interviewed higher percentage of people crossing for work (35.3%). It is important to note that most interviewees in Bad Radkersburg were Slovenes. When it comes to meeting friends or relatives, none of the respondents in Gornja Radgona cross the border for this reason (100%), whereas in Bad Radkersburg, 23.5% of respondents do cross the border to meet friends or relatives. For recreation or sport, 15.5% of respondents in Gornja Radgona cross the border for this purpose. In Bad Radkersburg, a very small percentage (2.9%) of respondents cross the border for recreation or sport. Overall, these results highlight that respondents in Gornja Radgona have an equal likelihood of crossing the border for shopping but are significantly less likely to cross for work or to meet friends or relatives compared to those in Bad Radkersburg. Additionally, crossing the border for recreation or sport is more common among respondents in Gornja Radgona than in Bad Radkersburg.

Considering that the interviews were conducted during working hours and included Slovenians working in Austria, which may have influenced the patterns, we further examined the reasons for border crossings **by nationality**. Austrians primarily cross the border for shopping (50%) and social activities, such as meeting friends or relatives (20.8%), with a smaller percentage crossing for work (8.3%). Slovenes, on the other hand, show a more balanced distribution between shopping (36.8%) and work (35.1%), with a notable portion also crossing for recreation or sport (8.8%). These differences indicate that while shopping is a significant reason for both nationalities, Austrians are more likely to cross for social and leisure activities, whereas Slovenes have a considerable focus on work-related crossings.

Perceptions of Motives for Crossing the Border

To determine if the reasons for crossing the border align with what people perceive as the actual motivations, we also compared the views of Slovenian and Austrian respondents on why they believe individuals from the other side visit their town. Both Slovenians and Austrians predominantly believe that people from opposite side cross the border for shopping: 50% of Slovenian respondents and 41% of Austrian respondents. Slovenians believe Austrians come for general shopping, purchasing cheaper goods, and accessing services like cosmetics or groceries that are perceived to be less expensive in Slovenia. Austrians believe Slovenians come shopping for goods that might be perceived as better quality or specific products not easily available in Slovenia.

In addition to shopping, recreational activities such as visiting cafes ("for a coffee and ice cream"), enjoying leisure activities, and using beauty and wellness services (spas, cosmetic services, hairdressers) are frequently perceived as motives for Austrians by Slovenians. They also noted that Austrians might be attracted to Slovenia for tourism and cultural experiences, such as exploring cultural sites or participating in events. On the other hand, Austrians frequently believe that Slovenians cross the border (beyond shopping) for work-related reasons. A significant number of responses also mentioned education and professional services, like going to specific stores or accessing services such as car maintenance or healthcare that they believe might be better or more specialized in Austria.

Language Understanding Issues

We analyzed language understanding issues by addressing two different aspects: proficiency of the neighboring country's language and difficulties in communicating with people across the border. Generally (as seen in Fig. 3) both nationalities think they have no communication issues. However, language proficiency analysis by nationality reveals that Austrians generally rate their Slovene language skills lower, with most describing their knowledge as "very poor" (25%) or "poor" (29.2%). In contrast, Slovenians show a broader range of ratings in regard to German, with the highest proportions indicating "good" (33.3%), "average" (24.6%), and "very good" (17.5%) proficiency. A smaller percentage of Slovenians rate their knowledge as "don't understand" (3.5%) or "very poor" (1.8%) compared to Austrians.

On the other hand, qualitative analysis of responses regarding difficulties in communicating with people across the border revealed that the "power of the language" (assuming that "German being more powerful than Slovene") is more important than proficiency of neighboring language (Fig. 4). The responses from Slovenians indicate a mix of experiences with communicating across the border. For them, good communication across the border means knowing German. Nearly half of the Slovenian respondents (47.4%) reported no issues with understanding German. A small percentage mentioned minor difficulties. Some Slovenians noted that they cannot communicate well, because people across the border don't speak Slovene. A few respondents highlighted that they could manage communication adequately ("I understand enough to be able to communicate") or have no problems at all. Additionally, some Slovenians expect Austrians to speak Slovenian when visiting Slovenia (" I want them to speak Slovenian when they come here because we always have to adapt to them, but they don't adapt to us").

Conversely, Austrians most commonly report about communication issues when people across the border don't speak their (German) language and not because they don't speak the language of Slovenia. A significant portion of Austrians (37.5%) therefore reported no language issues. Other responses highlighted the bilingual nature of the region, where many Slovenians can speak German ("No issues. Everyone can speak German"). A few Austrians mentioned facing specific challenges like younger people not speaking German.

The "other" group (including a Ukrainian and a Macedonian from Bad Radkersburg (Radgona), as well as two Serbians and one Croatian from Gornja Radgona) generally do not speak German, as indicated by responses such as "Austrians don't want to speak Slovene at all" or "I don't understand German at all."

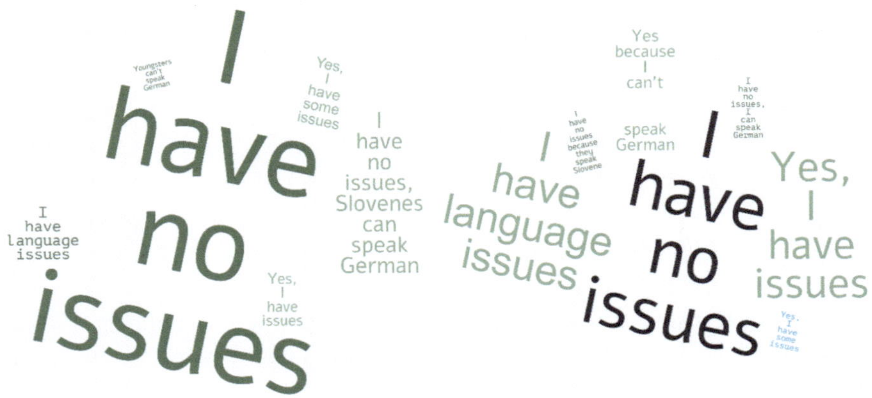

Fig. 4 Word cloud representing responses to the question "Do you experience any difficulties in communicating with residents on the other side of the border (e.g., language-related)?" Left: Austrians, Right: Slovenes

Conclusion and Discussion

This chapter focuses on two border territories: the towns of Gornja Radgona in Slovenia and the neighboring Bad Radkersburg (Radgona) in Austria, along with the Slovene German bilingual hinterland comprising five villages, known as the Radgonski kot (Radgona Corner) region in present-day Austria.

During the period of socialist Yugoslavia, the border with Austria was relatively open compared to other socialist states, allowing for some degree of cross-border cooperation. However, this cooperation underwent significant changes following Slovenia's transition to a market economy and particularly after its accession to the EU and the Schengen Area, which eventually resulted in the dismantling of hard border infrastructure. The introduction of Schengen facilitated even greater cross-border movement and collaboration, further integrating the region.

Therefore, in this study, we analyze the present-day cross-border dynamics in the areas of culture, economy, sports, and leisure, and how these interactions promote intercultural learning and bilingualism. We also examine the role of the billingual heritage of the Radgonski kot region, now part of Austria, within this context. Based on the research results, we discuss the four dimensions of our finding:

The Impact of Cross-Border Cooperation

Although there are varied socio-economic dynamics at play in cross-border interactions between Gornja Radgona and Bad Radkersburg (Radgona), the main motivations for crossing the border among Austrians are shopping and social activities, such as meeting friends or relatives, while Slovenes mainly cross for shopping and work, with some also crossing for recreation or sport, particularly on well-maintained cycling, jogging, and strolling routes. This study has also demonstrated the significant role that cross-border cultural and economic cooperation plays in fostering intercultural understanding, preserving linguistic heritage, and promoting bilingualism.

Changing Importance of the Linguistic Heritage of the Region

As Zupančič (2005) argues, national minorities have historically played a crucial role in facilitating communication between countries. However, with increased mobility reducing language and communication barriers between people from different cultural backgrounds, direct communication has become more prevalent. The "bridge" has been built, traffic flows freely, and minorities are starting to lose their role as connectors (Zupančič 2005). This trend is evident in our study as well, where although the Slovenian language holds significance in the region, German dominates due to its larger speaker base and stronger economic influence.

Key Findings from the Fieldwork with Students

Students experienced certain challenges in fieldwork, such as the difficulty in engaging participants and the impact of time and location on the diversity of respondents. Several students expressed that "*there were not enough people on the streets, and some did not want to participate*", and another mentioned, "*the area we were in didn't have much activity*". Despite these challenges, the fieldwork provided students with practical experience in data collection and analysis, contributing to their understanding of cross-border relations and the role of cultural and linguistic factors in these interactions. One student summarized this learning experience by stating, "*I learned about friendly relations between Slovenes and Austrians and how life in the cross-border area influences bilingualism or monolingualism.*"

Students observed that language plays a significant role in cross-border communication. Although the Slovenian language is important in the region, German often dominates, especially in economic contexts. One student commented, "*I learned that even near the border, many do not speak both languages, but there is still a strong connection between people*". This reflects a broader trend where direct communication across the border is becoming more prevalent, reducing the traditional role of minorities as cultural connectors.

The successful integration of GIS into education of undergraduate students has proven invaluable in designing, collecting and analyzing data that provides insights into the dynamics of linguistic heritage and cross-border activities.

Study Limitations and Future Research

Despite the valuable insights gained, the study had some limitations. The survey's timing during working hours limited the diversity of respondents, skewing the data towards older age groups and Slovenes who were presumably overrepresented on the Austrian side (due to their work commitments in Austria). Additionally, the reluctance of some individuals to participate and the language barriers encountered may have affected the breadth and depth of the data collected. These limitations should be considered when interpreting the study's findings and in planning future research efforts.

Future research should aim to better include members of the Slovene minority in the survey to explore more comprehensively the importance of the autochthonous minority in cross-cultural cooperation. By ensuring greater representation of this group, future studies could provide a more nuanced understanding of how the indigenous Slovene population contributes to and is affected by cross-border interactions, thereby enriching the discourse on the role of minority communities in fostering intercultural understanding and cooperation.

Acknowledgements This work was supported by the research program P5-0181, "Sociological Aspects of Sustainable Sociospatial and Manpower Development of Slovenia in Europe," financed by the Slovenian Research Agency. The authors would like to acknowledge the contributions of the 2023/2024 cohort of students who attended the Social and Cultural Geography course at the University of Maribor and collected the field responses. We also wish to express our gratitude to Sebastian Walcher and Susanne Weitlaner for their invaluable assistance in this study by familiarizing us with the activities of the Slovenian minority in Radgonski kot and for hosting our group at Pavlova hiša / Pavelhaus. Their support and insights into the work of the Artikel-VII-Kulturverein für Steiermark / Kulturno društvo Člen 7 za avstrijsko Štajersko, greatly contributed to the success of our research.

Literature and Sources

Bad Radkersburg (2024) Bad Radkersburg. Retrieved on 21 July 2024 from https://www.bad-radkersburg.gv.at/die-stadt/begruessung/

Bufon M (2004) Slovenia as a European contact area. Slovenia: a geographic overview, pp 21–26. https://zgs.zrc-sazu.si/Portals/8/Slo_Geo_Over/4.pdf

Gerfert S (2009) Cross-border cooperation; transforming borders. https://essay.utwente.nl/60149/1/BSc_S_Getfert.pdf

Gov.si (2024) Avstrijska državna pogodba po skoraj sedmih desetletjih od podpisa še vedno čaka na polno uresničitev. Retrieved 19 July 2024 from https://www.gov.si/novice/avstrijska-drzavna-pogodba-po-skoraj-sedmih-desetletjih-od-podpisa-se-vedno-caka-na-polno-uresnicitev/

Gov.si (2024a): Slovenci v Avstriji. Retrieved on 20 July 2024 from https://www.gov.si/teme/slovenci-v-avstriji/

Gostenčnik J, Kenda-Jež K, Kumin-Horvat M (2022) Ogrožena narečja v slovenskem jezikovnem prostoru. Jezikovni zapiski 28:75–87. https://www.ojs.zrc-sazu.si/jz/article/view/11269/11344%20, https://www.proquest.com/docview/2197982631?pq-origsite=gscholar&fromopenview=true&sourcetype=Dissertations%20&%20Theses

Haberl-Zemljič A (2004) Die Sprache im Dorf lassen: Festhalten und Aufge-ben der slowenischen Sprache in Radkersburg Umgebung, Graz: Artikel-VII-Kulturverein für Steiermark – Laafeld: Pavelhaus = Potrna: Pavlova hiša

Jesih B (1998) Slovenci Na Avstrijskem Štajerskem. Enciklopedija Slovenije 12:259–260

Klemenčič M (1978) Jezikovna struktura prebivalstva na z avtohtonim slovenskim prebivalstvom poseljenem območju avstrijske Štajerske od sredine 19. stoletja do leta 1971. ČZN 14:124–148. https://www.sistory.si/media/legacy/publikacije/7001-8000/7005/1978_1_Casopis_za_zgodovino_in_narodopisje.pdf#page=130

Klemenčič V (1994) Narodne manjšine kot element demografske in prostorske stvarnosti v alpsko-jadransko-panonskem prostoru. Geographica Slovenica 24:19–31. https://giam.zrc-sazu.si/sites/default/files/gs_clanki/GS_2401_019-031.pdf

Kladnik D, Pipan P (2008) Bay of Piran or Bay of Savudrija? An example of problematic treatment of geographical names. Acta Geographica Slovenica 48(1): 57–91. https://doi.org/10.3986/AGS48103

Klotz G (2019) Narodne manjšine kot dejavnik pri spodbujanju čezmejnega sodelovanja. Zamejski, obmejni, čezmejni Slovenci v Furlaniji - Julijski krajini deset let po vstopu Slovenije v šengensko območje, pp 29–34. https://www.slori.org/wp-content/uploads/2019/06/zbornik_schengen_def.pdf

Križman M (1987) Jezikovna sedanjost v Radgonskem kotu s socialno-lingvističnih vidikov. ČZN 23:173–215. https://www.sistory.si/media/legacy/publikacije/7001-8000/7031/1987_2_Casopis_za_zgodovino_in_narodopisje.pdf

Popis prebivalstva 2002 (2002) Prebivalstvo po maternem jeziku. Retrieved 20 July 2024 from https://www.stat.si/popis2002/si/rezultati/rezultati_red.asp?ter=SLO&st=55

Prunk J (2006) Nastanek republike Slovenije in njen razvoj. Retrieved 18 July 2024, from http://www.15let.gov.si/si/15-let-samostojnosti/nastanek-razvoj/

Rihtarič I (2026) Okraj Gornja Radgona 1945–1950. Univerza v Mariboru, doktorska disertacija. DKUM - Okraj Gornja Radgona 1945–1950

Robežnik N (2024) Določitev slovenske zahodne meje v 20. Stoletju. Retrieved 19 July 2024 from https://www.muzej-nz.si/izobrazevanje/interaktivne-vsebine/ucna-gradiva/dolocitev-slovenske-zahodne-meje-v-20-stoletju/

Stergar J (2001) Zamejski Slovenci. Enciklopedija Slovenije XV:60–62.

Svenšek A, Kosmač G, Kač M, Horvat K, Bergant IE (2020) Sto let koroškega plebiscita: zeleno veselje in bela žalost. Retrieved 19 July 2024, from https://www.rtvslo.si/mmcpodrobno/sto-let-koroskega-plebiscita-zeleno-veselje-in-bela-zalost/538476

Slovenija (2024) Zamejski Slovenci. Retrieved 19 July 2024 from https://eucbeniki.sio.si/geo9/2598/index5.html

Weitlaner S (2023) Slovensko kulturno delovanje na avstrijskem Štajerskem. Svetovi 1(1):121–126. https://journals.uni-lj.si/svetovi-worlds/article/view/11413/11946

Zorko Z (1989) Govor vasi Žetinci (Sicheldorf) v avstrijskem Radgonskem kotu. Slavistična revija 37:241–252. https://www.dlib.si/stream/URN:NBN:SI:DOC-RLPMMLJS/3869e697-f26d-4b02-bb90-6cf480d05fc0/PDF

Zupančič J (1997) Slovenci v Avstriji - število, način poselitve, struktura, identiteta. Geografski vestnik 69:115–138. https://zgs.zrc-sazu.si/Portals/8/Geografski_vestnik/Pred1999/GV_6901_117_140.pdf
Zupančič J (2005) Vloga in pomen narodnih manjšin v čezmejnem sodelovanju; Primer manjšin v Sloveniji in Slovencev v sosednjih državah. Razprave in gradivo - Inštitut za narodnostna vprašanja 47:26–41. https://www.dlib.si/stream/URN:NBN:SI:doc-2AT7D9LA/5964ccf1-c0bc-459d-aeef-bbc5aa5e5ddf/PDF
Zupančič J (2017) Slovene border landscape transformation. J Geogr Polit Soc 7(1):10–18. https://www.ceeol.com/search/article-detail?id=557057

Application of Geographic Information Systems in Social Science Teacher Training: Analysis of the Evaluation of Natural, Cultural, and Landscape Heritage in Depopulated Areas

José Antonio Mérida Donoso, Carlos Guallart Moreno, Alberto Serrano Andrés, and Lorenzo Mur Sangrá

Abstract Aware of the need for future teachers to acquire digital competencies in handling GIS as a key tool to analyze the evolution of natural, cultural, and landscape heritage, this article presents an intervention using GIS based on a comparison of aerial photographs in depopulated rural areas. The results highlight the high potential of such activities to foster students' digital and spatial competencies, document and analyze the state of heritage conservation over time, and assess potential threats through geospatial data analysis. This approach fosters meaningful learning through critical analysis, enhancing awareness and promoting an educational model that can be applied to other regions in Spain and Europe.

Resumen Conscientes de la necesidad de que los futuros docentes adquieran las competencias digitales de manejo de los SIG como herramienta clave para analizar la evolución del patrimonio natural, cultural y paisajístico, el siguiente artículo recoge una intervención con SIG basada en una comparación de fotografías aéreas en zonas rurales despobladas. Los resultados advierten del alto potencial de este tipo de actividades para fomentar las competencias digitales y espaciales del alumnado, documentar y analizar el estado de conservación del patrimonio a lo largo del tiempo y

J. A. M. Donoso (✉) · L. M. Sangrá
Faculty of Education, ARGOS Research Group, IUCA-Unizar, University of Zaragoza, Zaragoza, Spain
e-mail: jamerida@unizar.es

L. M. Sangrá
e-mail: lmsangra@unizar.es

A. S. Andrés
GEOT Research Group IUCA-Unizar, School of Tourism, University of Zaragoza, Zaragoza, Spain
e-mail: aserranoandres@unizar.es

C. G. Moreno
Faculty of Education, University of Zaragoza, Zaragoza, Spain
e-mail: cguallart@unizar.es

R. de Miguel González and P. Rivero Gracia (eds.), *Geospatial Technologies for Heritage Education*, Key Challenges in Geography,
https://doi.org/10.1007/978-3-031-96372-8_8

evaluar sus posibles amenazas desde el análisis de datos geoespaciales. De esta forma se genera un aprendizaje significativo mediante un análisis crítico que ayuda a una mayor sensibilización mediante un modelo educativo extrapolable a otras regiones de España y Europa.

Keywords Geospatial analysis · Heritage · Depopulation · Social science education · Teacher training

Palabras Clave Análisis geoespacial · Patrimonio · Despoblación · Educación en ciencias sociales · Formación del profesorado

Introduction

In recent decades, the use of Information and Communication Technologies, and more specifically Geographic Information Systems (GIS), has become established for the study of space, with a corresponding educational transposition (Boix et al. 2009). Their implementation in the classroom stems from the potential of GIS to surpass the limitations imposed by mere visual exploration, although it requires a basic level of knowledge to manage the information appropriately and maximize its utility (Boix and Olivella 2007). In particular, activities such as this one enable the analysis of changes over time in specific territories through visual comparisons that facilitate the development of geographical competencies. Within this framework, this study focuses on Spain's depopulated areas to address a heritage that tends to be largely neglected (Scazzosi 2018). As an open, negotiable, dialogical, transformative, and reflective concept, heritage (Fontal and Martínez 2017, p. 71) requires that when approaching rural spaces, the notions of debate and change must play a fundamental role (López-Estébanez et al. 2024), since these spaces are shaped by the values society assigns to them at any given time, which are subject to change (Baena Gallé 2016, p. 8).

In light of the above, we can agree that, in the educational field, GIS represents an indispensable tool for understanding and analyzing landscapes, as it integrates spatial data and attributes into a single system, allowing for the visualization and analysis of complex information in comprehensible and accessible ways (Boix and Olivella 2007; Torralba 2021). In fact, by creating thematic maps, digital terrain models, and spatial analysis, students can investigate the interactions between environmental, social, and economic factors that shape the landscape. In this way, the ability to overlay and analyze multiple layers of information within a spatial context fosters a deeper understanding of the processes affecting the environment (Mártinez and Molina 2015; Torralba 2021; De Migel et al. 2019).

Aerial photography, when used in combination with GIS, further amplifies this analytical capability, as aerial images provide a panoramic view of the landscape that reveals patterns and features that may not be evident from the ground (Sevilla and Rodríguez 2015). These images provide crucial visual data for analyzing changes in land use, erosion, urban expansion, and other geographic phenomena. Based on these premises, this paper presents the design of a practical methodology that enables the extraction of information from aerial photographs integrated into GIS. This methodology employs digital processing techniques to apply visual interpretation, aiming to stimulate debate based on the changes observed.

This methodology enhances the accuracy of spatial analysis by combining high-resolution aerial images with GIS, which allows for the detection of changes in land use and environmental patterns. Historical and contemporary images can be compared to track long-term changes, such as soil erosion or urban growth. However, the precision of the analysis depends on the quality and resolution of the images used, making careful data selection and validation important to ensure reliable results.

Objectives

1. Promote digital, spatial, and innovation competencies among teachers through the implementation of GIS.
2. Establish comparative frameworks using GIS to reflect changes in the landscape, aiming to document the conservation status of identified heritage and its potential changes and deterioration over time.
3. Evaluate threats through geospatial data analysis to identify risk areas in terms of erosion or deterioration of landscape heritage, thereby facilitating awareness and planning of possible conservation and risk management measures.
4. Promote education and awareness through GIS, including interactive maps and photographic galleries within an educational material model that can be extrapolated to other European regions.

Methodology

To promote the use of GIS in general, and ArcGIS Online in particular, within Spanish school classrooms, a proposal was sent for a classroom activity focusing on depopulated Spain, specifically La Ribagorza (Aragon) and its heritage, as a model for secondary and high school activities. The proposal consisted of comparing two rural areas on a map of the region, chosen by the students based on their interests or curiosity. The proposed historical landscape review is based on cartographic viewers like those from the National Geographic Institute (Instituto Geográfico Nacional) and its National Aerial Orthophotography (Plan Nacional de Ortofotografía Aérea) orthophoto comparator (Centro Nacional de Información Geográfica 2020), available for verification and use via the provided https://ggeomentoresesp.maps.arcgis.com/apps/webappviewer/index.html?id=115b84cb2fa9413eb42b3bc86401fed6. To observe the changes in their surroundings over the last 68 years, students filled out a form noting the changes perceived in the landscape from 1956 (American

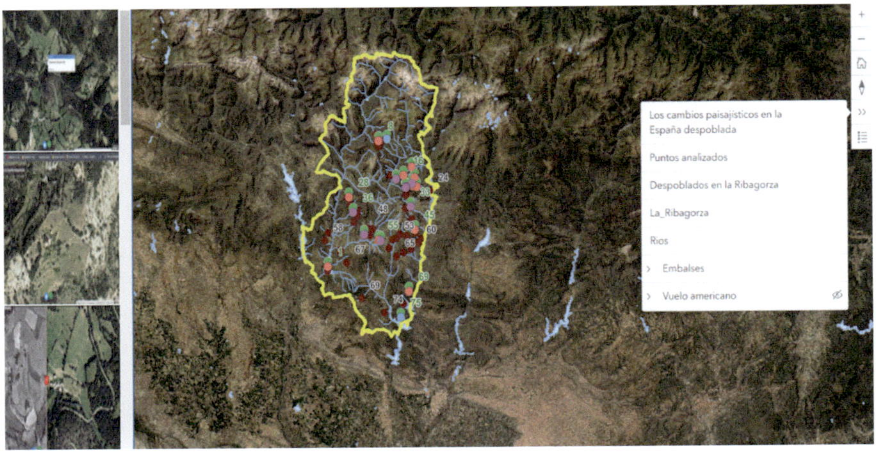

Fig. 1 Screenshot of the La Ribagorza region with screenshots of the points selected by the students on the left margin

flight) to the present (current orthophoto). This approach aimed to verify how the integration of aerial photographs into GIS analysis allows for direct observation of landscape changes over time, thereby facilitating the identification of trends and the formulation of hypotheses based on real data. On the website, students could switch between different periods instantly, maintaining various layers or filters on the map that enabled them to identify depopulated towns, rivers, and reservoirs, with the option to zoom in or out (Fig. 1).

To gather as much data as possible, participants were divided into two groups:

1. Teachers and future educators who are interested in implementing such practices in the classroom. This included students from the Faculty of Education at the University of Zaragoza enrolled in the Master's in Secondary Education, specializing in Geography and History (2023–2024) and active teachers, totaling 30 participants. The goal was to analyze their feedback, identify potential improvements and difficulties, and gauge their expressed interest in conducting this comparison in class.
2. A practical implementation in the classroom to evaluate perceptions and learning in real-time. A session was conducted at the Joaquín Costa Institute in Cariñena with students aged 11–15 (21 students). This institute serves secondary education students from surrounding municipalities and some from the Campo de Daroca region (Herrera and Villar de los Navarros), some of whom face the challenge of depopulation. This approach aimed to achieve a dual purpose: first, for students from rural backgrounds to analyze how depopulation influences the current landscape configuration; second, for the Geography and History teachers to recognize the educational potential of GIS in reflecting on their territorial reality.

The intervention was designed to be completed in 10–15 min, either in the classroom or outside, but in practice with secondary students, a full session was conducted to allow maximum freedom for the teacher. The activity focused on the Ribagorza region and its heritage in depopulated areas to address its heritage, the possible changes and deterioration suffered over time, and to identify areas of risk or deterioration of landscape heritage.

Once a point of interest was selected, students could capture an image to make a comparison and identify changes in the analyzed environment according to the following categories:

- Urbanization
- Agricultural Changes
- Temporary Changes
- Infrastructure
- Abandonment/Neglect
- Environmental Changes.

Through the qualitative and quantitative analysis of the collected data, we expect students to gain a deeper understanding of how depopulation has impacted their local landscape. In past sessions, we have noticed that when students discuss the changes identified in different categories, they not only engage with the material but also start reflecting on their environment and the challenges it faces. For instance, during the analysis of earlier images, some students were surprised to discover the impacts and changes they observed in the region. These discussions encourage meaningful dialogue, allowing both students and teachers to connect these changes to broader socioeconomic trends. By the end of this process, we believe students will recognize the importance of preserving their landscape heritage and feel motivated to get involved in conservation initiatives within their communities.

Results

Geographical Analysis of the Location of Points on the Map

The analysis of the chosen points reveals that some generated more interest, whether due to personal reasons or because they reflected significant changes in the landscape. This might be linked to the diversity of transformations observed or the accessibility of those locations for the respondents.

Point 55 was the most selected, with 4 mentions.

Points 8 and 18 were chosen 3 times each.

Points 7, 19, and 28 were selected 2 times each.

Other points, such as 1, 16, 17, 24, 33, 36, 53, 61, 67, 69, 74, and 75, were chosen only once (Fig. 2).

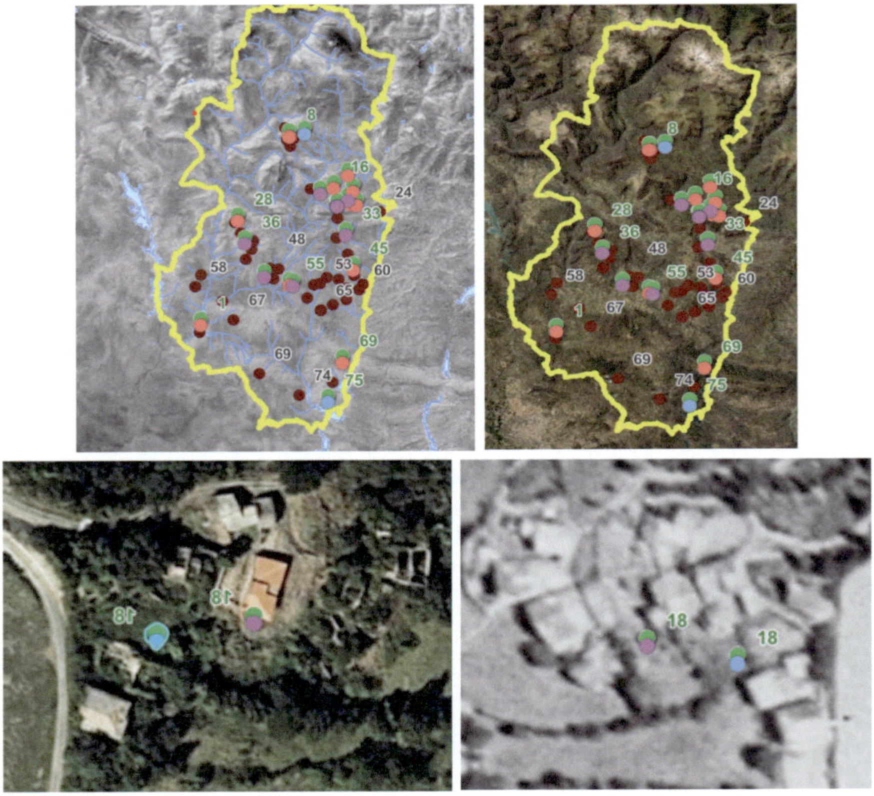

Fig. 2 Images of La Ribagorza without zoom and with zoom from point 18

As can be seen, some points have a higher level of interest and are likely to show more notable changes in the landscape. A more detailed analysis of the location of the selected points in La Ribagorza reveals relevant aspects of their distribution and changes.

General Distribution

The points are mainly dispersed in the central and northeastern parts of La Ribagorza, with a notable concentration near river networks and low-altitude areas, such as valleys and flatlands, which are more accessible and have historically been more populated. This concentration in the northeast might relate to proximity to urban centers or areas with a higher population density and economic activity.

Relationship with the Hydrographic Network

Many points are located near rivers and streams, such as the Ésera River and its tributaries. This proximity suggests that landscape analyses have particularly emphasized areas where water plays a key role, whether in terms of water resources, flood risk, or as a factor influencing land use. These water-adjacent areas are vital for managing water resources and conserving riparian ecosystems, highlighting the importance of protecting and sustainably managing these zones.

Topography and Accessibility

The points are also distributed along slopes and valley bottoms, reflecting a particular focus on changes in agricultural and forested landscapes typical of these areas. Accessibility may influence the selection of analysis points; more remote or difficult-to-reach areas, such as higher mountains, have fewer points, limiting direct observation of landscape changes in those regions.

Urbanization and Depopulation

Points located near towns and population centers reflect urban changes, such as the expansion or reduction of human habitats, as well as the direct influences of urbanization or abandonment of structures. The dispersion and density of these points could correlate with rural depopulation phenomena, a challenge in territorial management for many mountainous areas in Huesca.

Natural and Cultural Heritage

La Ribagorza is rich in cultural and natural heritage, including natural parks and historical sites. The location of these places indicates an interest in analyzing how these heritages are being affected by environmental and human changes, such as landscape erosion or alteration of traditional routes. This focus is crucial for developing preservation and promotion policies for heritage in the region.

Implications for Territorial Management

The distribution of the points provides a clear guide on where landscape analysis efforts are concentrated in La Ribagorza. These points highlight key areas for territorial management, including the conservation of natural resources, adaptation to climate change, and land use planning to prevent uncontrolled expansion or degradation of landscapes.

Geographical Analysis of Changes in the Analyzed Environment

The changes observed in the analyzed environment are categorized as follows (Fig. 3):

- Urbanization (44 mentions): This category includes changes related to the development of urban centers, expansion of towns, and the construction of houses and other buildings.
- Agricultural Changes (35 mentions): Refers to transformations in crop fields, the expansion of agricultural areas, and other shifts in land use.
- Temporal Changes (33 mentions): Includes references to changes over time, such as comparisons between past and present (e.g., "now," "current," "before," "1956").
- Infrastructure (17 mentions): Related to modifications in communication networks, roads, and other infrastructures.
- Abandonment/Neglect (15 mentions): Reflects deterioration, abandonment, or disappearance of elements in the landscape.
- Environmental Changes (14 mentions): Includes alterations in vegetation, trees, water bodies, and other natural elements of the landscape.

The bar graph illustrates the number of observations for each change category (such as urbanization, agricultural changes, infrastructure, abandonment, and environmental changes) at the most frequently selected points. Analyzing this data reveals that the most commonly observed changes are related to urbanization and agricultural shifts, followed by variations over time and modifications in infrastructure.

The most common themes in the responses regarding changes in the analyzed environment included terms related to transformations in land use and landscape structure. The frequently used words were as follows:

- "Fields": Indicates changes in rural areas, possibly related to agriculture or the expansion of cultivated land.
- "Nucleus" and "Town": Refers to changes in urban areas or population centers, such as town expansions or infrastructure changes.
- "Crop": Suggests modifications in cultivated areas, whether in terms of size or types of crops present.

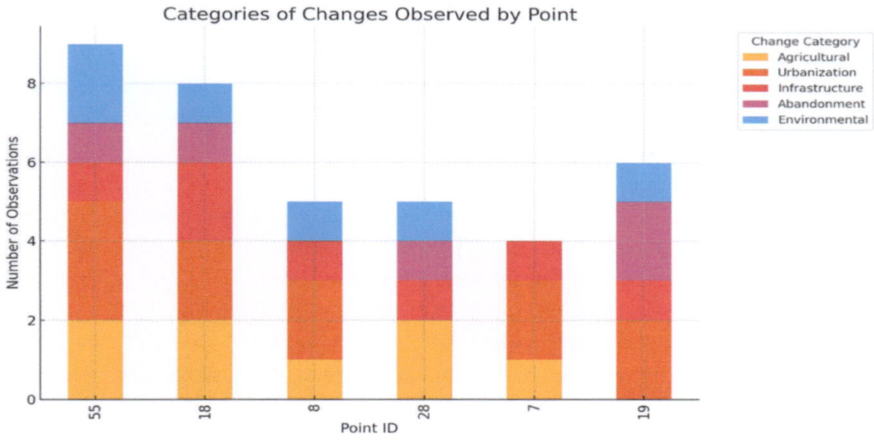

Fig. 3 Graphical visualization of change categories observed by each point

- "Routes" and "Paths": Indicates changes in transport infrastructure, such as the appearance or disappearance of roads.
- "Abandonment": Signals the deterioration or neglect of certain areas, possibly linked to a decline in human or agricultural activity.
- "Disappeared": Refers to the loss of landscape elements, such as structures, cultivated areas, or vegetation.

These terms imply that the main changes observed in the landscape are related to shifts in land use, the expansion or contraction of inhabited and cultivated areas, and changes in transport infrastructure, reflecting a process of humanization or abandonment of the environment. The results regarding types of change in the analyzed environment indicate urbanization (44 mentions), agricultural changes (35 mentions), temporal changes (33 mentions), infrastructure (17 mentions), abandonment/neglect (15 mentions), and environmental changes (14 mentions). As seen, the results show that the most frequently observed changes are linked to urbanization and agricultural shifts, followed by variations over time and modifications in infrastructure.

Finally, the analysis of the feedback on the utility of GIS for landscape analysis reveals:

- "Very Useful": The vast majority of respondents (22 teachers) consider GIS to be very useful for landscape analysis. This indicates a positive perception and strong appreciation of GIS as an effective tool for this type of analysis.
- "Fairly Useful": 7 respondents believe GIS is useful, reflecting a favorable view, albeit with slightly less enthusiasm compared to the previous group.
- "Not Very Useful": Only 1 respondent thought GIS was not very useful for landscape analysis, suggesting a negative perception, although this is a minority opinion.

Overall, the results indicate a positive consensus regarding the usefulness of GIS for landscape analysis, with most participants acknowledging its value and effectiveness in this context. This is particularly true for both current and future teachers, who highlighted GIS's significance for studying the landscape heritage of rural areas, taking into account aspects like student interest and proximity. Participants emphasized that the tool facilitates the visualization of phenomena such as landscape heritage, depopulation, and rural heterogeneity—essential aspects for understanding the contemporary rural geographical reality (Bandrés and Azón 2021). In this context, both current and future teachers underscored GIS's high potential to shed light on the causes and consequences of depopulation in Spain.

Additionally, the qualitative analysis of feedback from students at IES Joaquín Costa de Cariñena shows that this practice generated more meaningful learning experiences. This aligns with similar research findings (Palmer 2013). The choice of spaces to explore through aerial photography provided a sense of virtual excursion or, as some may prefer, a sentimental journey through a part of Spanish geography. This freedom for students to select comparison spaces and combine various layers of geographical information translates into a deeper understanding of the complex interplay between physical and human factors shaping landscapes, fostering critical and creative thinking. Consequently, participating students were able to effectively identify and analyze changes in the landscape over time, demonstrating how the visual comparison of aerial photographs aids in grasping complex geographical concepts, such as the consequences of rural exodus and the transformation of rural spaces.

Discussion

As observed in the practice, the points on the map represent locations where changes in the landscape have been observed and analyzed, which have direct implications for territorial management. In this regard, several key aspects emerge regarding how these points can influence territory management:

1. Identification of Critical Areas: The points concentrated in certain areas, especially near rivers and valleys, indicate zones of high interest or concern. These areas may be experiencing significant changes such as urbanization, agricultural expansion, or erosion. Knowing these locations allows land managers to prioritize interventions and resources at these critical points.
2. Urban and Rural Planning: Points that show a high degree of urbanization or abandonment can guide decisions regarding urban and rural planning. For example, in areas where uncontrolled urban expansion is observed, authorities might implement stricter zoning policies or develop adequate infrastructure to mitigate environmental impacts.

3. Environmental Conservation: Points located in regions of high biodiversity or near water bodies suggest the need for conservation strategies. For instance, if points indicate negative changes in vegetation or loss of natural habitats, managers can develop reforestation programs or watershed protection initiatives.
4. Water Resource Management: The proximity of many points to river networks highlights the importance of properly managing water resources. This may include protecting aquifer recharge areas, preventing water contamination, and managing flooding in vulnerable areas.
5. Climate Change Adaptation: Changes observed at these points, such as the abandonment of agricultural areas or urban expansion, may also be linked to climate phenomena. Understanding the location and nature of these changes can aid in developing climate change adaptation plans, such as creating green barriers or planning resilient infrastructure.
6. Sustainable Development: Information derived from these points can support sustainable development initiatives, promoting a balance between human development and environmental conservation. This includes advocating for sustainable agricultural practices, regulating land use, and preserving cultural landscapes.

In summary, the points on the map are fundamental for understanding the local dynamics of the landscape and are therefore crucial tools for decision-making in territorial management. They enable managers to prioritize actions, tailor policies to the specific needs of the territory, and promote more efficient and sustainable land use. Consequently, we can assert that through this activity, students create, plan, organize, and articulate new knowledge with a higher level of cognitive complexity.

Existing literature on the use of geo-media and GIS in educational settings emphasizes the importance of generating innovation in the teaching of the discipline, highlighting the need to foster active and inductive methodologies, discovery learning (learning by doing), and ultimately, autonomous, critical, and functional learning (De Miguel 2013a, b; De Miguel and Sebastián 2022; Kolvoord 2012; Milson et al. 2012) that respond to these types of proposals.

In this context, for many years, the use of textbooks as a predominant and fundamental tool in the teaching and learning process of Geography, rather than serving as a supplementary resource, has tended to promote a traditional, lecture-based, fragmented, and encyclopedic approach to teaching. Despite advances in didactic resources outlined in Geography textbooks for Secondary Education and Baccalaureate (De Miguel 2013b), this reality partly accounts for Spain's distance from the necessary didactic innovation implemented in other countries (Buzo-Sánchez et al. 2022). In the case of depopulated Spain, the absence of rural geography in the curriculum and textbooks addressing significant changes in the rural environment over recent decades further complicates matters (Quintá et al. 2018). In this sense, developing this activity provides teachers with a new tool to enhance the significance of learning derived from their teaching activities (Zapettini et al. 2008).

The teachers responsible for conducting the activity in the classroom recognized the importance of applying GIS to renew and enrich geography education, expressing interest in carrying out similar exercises focused on their territorial context. However, the lack of teacher training in the use of such tools (Albert and Nieto 2014) poses a barrier that must be addressed to ensure that the development of these workshops is not a one-time event but a consistent part of the academic curriculum.

Thus, it is evident that GIS not only represents a technological tool filled with possibilities for classroom use but also acts as a catalyst for motivation, engagement, and collaboration to understand and solve problems that, from the specific, affect the global (Milson et al. 2012). In other words, it promotes thinking globally from the local perspective, leading to actions and considerations that are "glocal." Ultimately, this perspective fosters a geography of "social value" by offering a more dynamic view of the environment and a sense of belonging through analytical strategies that impact awareness of the social dimension of heritage, the sustainable connections between people and their surroundings, and the consequences of its deterioration. In this way, by aligning with a reformulation of the landscape, it enables education in the landscape rather than merely studying it (Busquets 2010; Lestegás et al. 2024), emphasizing the role that human action has played and continues to play, interpreting it as a social product or collective creation.

Notably, reinforcing its historical and cultural dimension highlights human actions, generating landscapes that evoke admiration, trivialization, or indifference (Busquets 2010), and placing it in a constant process of construction (Liceras 2013). Consequently, following proposals that contribute to developing civic awareness in students, enabling them to address various societal challenges (Sebastián-López and de Miguel-González 2017: 244), there is an opportunity to use digital platforms to revitalize these connections and help form a "glocal community" capable of recognizing the emotional ties people have with their environment, regardless of their physical location. An illustrative example of this social dimension is found in the memories of communities displaced by dam construction in Spain. These "drowned memories" reflect the lives of those forced to leave behind villages, land, and livelihoods, voices that were never consulted or heard (Marcos and Fernández 2024). Incorporating such perspectives into classroom work is not only educational but also an ethical gesture, recovering testimonies that deepen understanding of rural change and its human consequences. Such integration ultimately strengthens democratic awareness and a shared sense of heritage.

Conclusions

In light of the results obtained and to synthesize the possible aspects that facilitate improvements in learning processes, we understand that practices like the present one highlight three fundamental points or axes, namely:

- The need to promote the use of GIS and new technologies as instruments in educational processes and as elements for accessing landscape heritage.
- The creation of specific materials that facilitate the use and access to this heritage in the classroom.

- The promotion of collaboration between schools and/or educational institutions and universities to facilitate training, dialogue, access, and the creation of common programs based on synergy and the integration of educational institutions and universities.

As we have observed, the combined use of GIS and aerial photography in education not only enhances the understanding of geospatial concepts but also develops valuable technical skills. Students learn to operate advanced software, interpret spatial data, and apply analytical methods that are fundamental in the field of geography and many other disciplines. Furthermore, this integrative approach fosters critical thinking and the ability to solve complex problems, preparing students to face challenges in the professional realm and in scientific research. Thus, GIS and aerial photography consolidate as key tools in the training of future geography experts, contributing to a more dynamic and relevant education in landscape analysis.

The educational transposition of the rural heterogeneity of Spain and its landscape requires a deeper study in terms of methodology and analysis that allows for greater awareness and sensitivity to be brought into the classroom, as well as increased projection, protection, and promotion (Scazzosi 2018). Applications like the one presented allow for attention to the integrity of the landscape as a key identity marker, as noted by UNESCO (Gullino and Larcher 2013) amid the deterioration affecting depopulated rural Spain, in connection with demographic decline and the consequent abandonment of traditional activities such as agriculture and livestock farming. This abandonment leads to the loss of cultural landscapes that progressively deteriorate, including terraced fields, rural paths, or historic buildings. Similarly, it fosters soil erosion processes, uncontrolled vegetation growth, and an increased risk of fires, which ultimately affects not only the physical landscape but the entire social and cultural heritage linked to traditions and ways of life associated with the land. In summary, the disconnection between inhabitants and their environment alters a whole natural and cultural equilibrium and, therefore, a landscape that is a fundamental resource.

Reversing this situation requires emphasizing the importance of valuing landscape heritage in education as a basis for revitalizing these areas through awareness of their significance. GIS can facilitate the establishment of this heritage as a pillar upon which educational efforts should be built, necessitating certain changes in school organizations such as promoting collaborative work, improving teacher training in both specific technical aspects and didactic, pedagogical, and even cultural dimensions, and ultimately providing greater weight as a mediator of landscape heritage so that it feels a part of it and, therefore, plays a leading role in its conservation and dissemination within the educational sphere.

In conclusion, we understand that practices such as the present one demonstrate the opportunity that GIS offers for providing research-based learning about current events using real-world data, such as that related to depopulated Spain, in an environment of awareness, consciousness, and problem-solving driven by technology. The results show that the use of aerial photographs as a model for comparisons over time serves as a primary tool for generating participatory learning processes, monitoring heritage, and engaging the educational community in the collection, analysis, and representation of geographic data or information, promoting collaborative approaches to information, awareness, and conservation.

References

Albert JM, Nieto G (2014) Las TIG en la Educación Secundaria. La formación previa del profesorado. In: XVI Congreso Nacional de Tecnologías de la Información Geográfica. Available at: http://hdl.handle.net/10045/46767 (accessed 25 August 2025)

Baena Gallé JM (2016) Patrimonio y educación obligatoria. Deseos y realidades. UNES Universidad, escuela y sociedad, no 1, pp 6–20. Available at: https://revistaseug.ugr.es/index.php/revistaunes/article/view/12147 (accessed 25 August 2025)

Bandrés E, Azón V (2021) La despoblación de la España interior. Madrid: Funcas

Boix G, Olivella R (2007) Los Sistemas de Información Geográfica (SIG) aplicados a la educación. El proyecto PESIG (Portal Educativo en SIG). In Marrón MJ and Salom J, Las competencias geográficas para la educación ciudadana. Las competencias geográficas para la educación ciudadana pp. 23–32

Boix G, Olivella R, Sitjar J (2009) Los Sistemas de Información Geográfica en las aulas de educación secundaria. Geografía y Sistemas de Información Geográfica 1:17–36

Busquets Fàbregas J (2010) La educación en paisaje: una oportunidad para la escuela. Íber. Didáctica de las Ciencias Sociales, Geografía e Historia 65:7–16

Buzo-Sánchez IJ, Mínguez C, De Lázaro-Torres ML (2022) Expert perspectives on GIS use in Spanish geographic education. Int J Digit Earth 15(1):1205–1219. https://doi.org/10.1080/17538947.2022.2096131

Centro Nacional de Información Geográfica (2020) Instituto Geográfico Nacional (IGN). Comparador de ortofotos PNOA. Available at: https://visualizadores.ign.es/comparador_pnoa/ (accessed 26 August 2025)

De Miguel R (2013a) Aprendizaje por descubrimiento, enseñanza activa y geoinformación: hacia una didáctica de la geografía innovadora. Didáctica Geográfica 14:17–36. Available at: https://didacticageografica.age-geografia.es/ojs/index.php/didacticageografica/article/view/230 (accessed 25 August 2025)

De Miguel R (2013b) Geoinformación e innovación en la enseñanza-aprendizaje de la geografía: un reto pendiente en los libros de texto de secundaria. Didáctica de las ciencias experimentales y sociales 27:67–90. https://doi.org/10.7203/DCES.27.2344

De Miguel R, Sebastián M (2022) Competencias digitales y educación geográfica digital: los sistemas de información geográfica y la geoinformación. In Olcina J, Morote AF (Coord) La enseñanza de la geografía en el siglo XXI. Universidad de Alicante, pp 147–164

Fontal O, Martínez M (2017) Evaluación de programas educativos sobre Patrimonio Cultural Inmaterial. Estudios Pedagógicos (Valdivia) 43(4):69–89. https://doi.org/10.4067/S0718-07052017000400004

Gullino P, Larcher F (2013) Integrity in UNESCO world heritage sites. A comparative study for rural landscapes. J Cultural Heritage 14(5):389–395. https://doi.org/10.1016/j.culher.2012.10.005

Kolvoord B (2012) Integrating geospatial technologies and secondary student projects: the geospatial semester. Didáctica Geográfica 13:57–67. Avaliable at: https://didacticageografica.age-geografia.es/ojs/index.php/didacticageografica/article/view/96 (accessed 25 August 2025)

Lestegás FR, Rodríguez TR, Quintá FXA, Arce XCM (2024) Del estudio del paisaje a la educación en paisaje: Investigación y formación del profesorado de educación primaria (6–12 años). Agua y Territorio/water Landscape 23:193–206. https://doi.org/10.17561/at.23.7200

Liceras RA (2013) Didáctica del paisaje. Lo que es, lo que representa, cómo se vive. Íber. Didáctica de las Ciencias Sociales, Geografía e Historia 74:85–93

López-Estébanez N, Molina-Holgado P, Allende Álvarez F (2024) Preserving the values of Mediterranean enclosed fields with dry stone walls, an example of vulnerable natural and rural heritage. Heritage 7(2):844–872. https://doi.org/10.3390/heritage7020041

Marcos J, Fernández MªA (2024) Memorias ahogadas. Logroño: Pepitas editorial

Martínez LC, Molina I (2015) Docencia e investigación a través del análisis y la fotointerpretación del territorio. Los paisajes del Guadarrama (Segovia). Eria 98:309–326. https://doi.org/10.17811/er.98.2015.309-326

Milson AJ, Demirci A, Kerski JJ (Eds) (2012) International perspectives on teaching and learning with GIS in secondary schools, vol 56. Springer: New York. https://doi.org/10.1007/978-94-007-2120-3

De Miguel González R, Donert K, Koutsopoulos K (Eds) (2019) Geospatial technologies in geography education. Geogr Notebooks 2(2):105–109. Geospatial technologies in geography education. Springer International Publishing

Palmer RE (2013) Learning geomorphology using aerial photography in a web-facilitated class. Rev Int Geogr Educ Online (RIGEO) 3(2):118–137. Available at: https://dergipark.org.tr/en/pub/rigeo/issue/11182/133608 (accessed 25 August 2025)

Quintá FXA, Lestegás FR, Arce XCM (2018) La olvidada geografía rural en el currículo y manuales de la educación secundaria. REIDICS. Revista De Investigación En Didáctica De Las Ciencias Sociales 3:4–19. https://doi.org/10.17398/2531-0968.03.4

Scazzosi L (2018) Rural landscape as heritage: Reasons for and implications of principles concerning rural landscapes as heritage ICOMOS-IFLA 2017. Built Heritage 2:39–52. https://doi.org/10.1186/BF03545709

Sebastián M, de Miguel R (2017) Educación Geográfica 2020: Iberpix y Collector for ArcGis como recursos didácticos para el aprendizaje del espacio. Didáctica Geográfica 18:231–246. Available at: https://didacticageografica.age-geografia.es/ojs/index.php/didacticageografica/article/view/391/364 (accessed 25 August 2025)

Sevilla J, Rodríguez C (2015) La dinámica reciente del paisaje del puerto de Leitariegos (montaña occidental asturleonesa) a través de la fotografía aérea. Eria 98:241–274. https://doi.org/10.17811/er.98.2015.241-274

Torralba A (2021) Experiencia virtual de interpretación de cambios en el paisaje en la formación inicial de docentes de Educación Infantil. In Villalustre L, Cueli M (Coord) Avances y desafíos para la transformación educativa. Servicio de Publicaciones de la Universidad de Oviedo, pp 323–333

Zapettini MC, Zilio CK, Lertora LJ, Carut CB, Car NM (2008) Los Sistemas de Información Geográfica -SIG- en la enseñanza de la geografía. Tiempo y Espacio 18(21):94–112. https://doi.org/10.22320/rte.vi21.1742

The Hidden Heritage: Archaeology, Geophysics and GIS at the Service of Heritage Education in the Classrooms

Irene Ortiz Nieto-Márquez and Javier Vallés Iriso

Abstract Heritage is not always visible, particularly in archaeology, where much of it lies hidden underground. Often, archaeological remains are discovered accidentally and destructively, with governments sometimes overseeing their documentation. However, land surveying is crucial for proper documentation and interpretation. Using non-invasive or non-destructive techniques, as recommended by UNESCO, archaeologists can document hidden heritage. Geophysical surveys like ground-penetrating radar (GPR), alongside remote sensing and Geographic Information Systems (GIS), allow us to reveal entire buried cities, pipelines, defensive systems, and burial sites. Specialists and educators must document and share this historical heritage to ensure society values and preserve it. Teachers play a vital role in Heritage Education. Natural heritage, such as mines, and cultural heritage, like archaeological sites, can be used as educational resources in classrooms to teach History, Art History, and Geography. Incorporating archaeology encourages students to develop critical and reflective thinking. Current educational legislation and curriculum in Spain promote ICT tools, making GIS an ideal way to present hidden heritage through surveys and spatial analysis. By engaging students in archaeological research processes, they can explore new sites, solve historical questions, and develop teamwork and respect for heritage conservation.

Keywords Heritage · Archaeology · Geophysics · GIS · Active methodology · Critical thinking

I. O. Nieto-Márquez (✉)
Faculty of Education, Complutense University of Madrid, Madrid, Spain
e-mail: ireort01@ucm.es

J. V. Iriso
CAI of Archaeometry and Archaeological Analysis, Complutense University of Madrid, Madrid, Spain
e-mail: j.valles@ucm.es

R. de Miguel González and P. Rivero Gracia (eds.), *Geospatial Technologies for Heritage Education*, Key Challenges in Geography,
https://doi.org/10.1007/978-3-031-96372-8_9

Introduction

Nowadays, new educational systems in Spain increasingly focus on competency-based learning. This means that students, beyond acquiring knowledge, must know how to apply it in everyday situations (LOMLOE 2020).

This goal is achieved by changing the classroom methodology, shifting from a one-way teaching system—where students play a passive role as mere recipients of the teacher's knowledge—to a multidirectional learning system, where students actively participate in their own learning process, and the teacher acts as a guide and support.

Traditionally, experimental sciences and mathematics have been easier to apply in practical, everyday classroom situations. However, social sciences, often seen as more abstract and theoretical, have been relegated to traditional teaching methods, with lectures that are largely expository and reliant on memorization.

Within the realm of social sciences, we find Geography, History, Art History, Civic Education, and Heritage Education. These subjects are interconnected and can be studied in a transdisciplinary way. For example, we cannot study History without knowing where events took place (Geography), what was created during that period (Art History), what society was like at the time (Civic Education), and finally, how all of this can contribute to shaping us as critical and reflective citizens (Heritage Education).

Observing this, we must find a way to apply this knowledge to practical, everyday situations in which students can see themselves reflected and become active agents in their learning process, especially in the context of Heritage Education.

In this regard, Archaeology can be of great use to us. This is the science that studies past societies through the analysis and study of their material remains and artifacts (Renfrew and Bahn 1993). Archaeology is not only responsible for studying human history but also for making history, as the processes of surveying, excavation, and post-excavation analysis create narratives that reshape our understanding of past societies.

This inherently practical discipline, in addition to utilizing experimental sciences and mathematics, is fundamentally rooted in the social sciences (as well as geology). Therefore, we consider it a potential solution to bring social sciences to students in a practical and competency-based manner.

Objectives

The main objective of this work is to present a teaching proposal to carry out heritage education in the classroom in a competency-based and practical way. To achieve this, we will use archaeology and non-invasive surveying processes to identify possible underlying remains, invisible to society, and their subsequent interpretation through Geographic Information Systems (GIS) in the classroom.

One of the key objectives of Heritage Education is to educate by, for, and through Heritage. As stated in Article 4 of UNESCO's "Convention Concerning the Protection of the World Cultural and Natural Heritage," (1972) it is recognized that "the duty of ensuring the identification, protection, conservation, presentation and transmission to future generations of the cultural and natural heritage…" lies with us. This excerpt highlights two important concepts:

- The identification, protection, and conservation of heritage through non-invasive and non-destructive documentation techniques.
- The transmission of heritage to future generations through Heritage Education.

Since the origin of the discipline, archaeology has always been linked to excavation activities as the only process for obtaining data about past societies and their material remains (Micó Pérez et al. 1998; Ruiz Zapatero 2013). These are destructive processes that directly interfere with archaeological remains and their contexts.

Thus, archaeological excavation is an invasive method of obtaining information, and considering that heritage is limited and not infinite, we must minimize these activities in favor of non-invasive and non-destructive techniques.

For this reason, we must move beyond the idea that excavation is the only way to obtain information, as today we have numerous non-invasive techniques dedicated to the discovery of archaeological sites (Dietz et al. 2011).

Non-invasive archaeology is based on the application of new techniques from geophysics and remote sensing to detect potential archaeological sites (Faltýnová et al. 2015). These techniques are interpretive, making multidisciplinary work essential for accurate data interpretation.

These non-destructive techniques are also key to passing heritage on to future generations. If excavation practices are the only way to obtain data, there will come a time when we have no heritage left to pass on.

The dissemination and success of non-invasive archaeology is closely tied to Heritage Education. If society is not aware of its heritage, it will not be possible to preserve, care for, and transmit it: "The educational perspective on heritage is essential for it to be known, understood, valued, and transmitted by the society that receives and inherits it" (Fontal Merillas 2006).

According to the author, education serves as a bridge capable of fostering awareness, promoting appreciation, and enabling the transmission of values, facilitating mediation between individuals and their heritage-based identities. Through heritage references, it is possible to enhance reflective knowledge of reality, regardless of whether the objective is heritage conservation and appreciation (Estepa Jiménez and Cuenca López 2006).

Methodology

The proposed methodology to be used in the classroom will be **Problem-Based Learning (PBL)**, aimed at having students seek a solution using non-invasive surveying methods and GIS to document archaeological heritage without destroying it.

The application of this methodology in the classroom will follow the steps outlined by Eggen and Kauchak (2015). In the first stage, a question will be identified, presenting the central issue of the project: How can we document heritage without destroying it? In the second stage, students will generate hypotheses to solve the problem. Third, information will be gathered on non-destructive techniques in archaeology. Afterwards, students will evaluate the previously proposed hypotheses and conduct a practical application with the provided examples. To conclude the fifth stage, where the results obtained are generalized, and reflection on the process and evaluation takes place. In *Annex*, the instructional design of the project can be seen in more detail.

This methodology has been proposed because, as mentioned earlier, social sciences have traditionally been associated with passive teaching methods centred on memorization. PBL is ideal because it focuses on a real problem, organizing the content and skills students need to learn around it. Learning becomes student-centred, engaging both cognitive and affective processes to investigate and solve the problem. Students are asked to take responsibility for identifying learning objectives, planning the collection and processing of information, and reflecting on their understanding and learning. Collaborative work is essential, as is teacher scaffolding, whose role is to moderate and guide reasoning processes, information searches, and the formulation of relevant questions (Marra et al. 2014).

At the disciplinary level, to carry out this proposal, we have used Archaeology, Geophysics, and GIS to bring Heritage Education into the classroom.

The choice of **archaeology** as one of the disciplines to be implemented in this proposal is supported by the multiple empirical benefits it provides to both students and teachers.

Numerous publications have highlighted the advantages of using archaeological projects in classrooms, as they help develop educational values (Ayan Vilá et al. 2009; Castillo Lozano 2017; González Marcén 2010; Ruiz Zapatero 2010; Vicent et al. 2015) such as the management of sources, supporting the development of interdisciplinary and comprehensive teaching, promoting cooperative learning, instilling values such as respect and understanding of cultural diversity, and helping educate about the respect for cultural and natural heritage.

Additionally, as a scientific discipline, archaeology can foster scientific questioning, dialogue, and critical thinking, contributing to the scientific development of students (Montoya Martínez and Egea Vivancos 2021; Vicent et al. 2015). In this way, it allows an approach to scientific research methodology and helps overcome proposed problems, breaking away from purely memoristic and expository teaching methods in favor of a more active and meaningful one.

Finally, at a competency level, archaeology has proven to be a highly useful and applicable tool in classrooms, from early education to the university level. It involves multidisciplinary learning, opening up a wide range of competencies (basic, key, specific, etc.) (Egea and Arias 2013; Henson 2004; Ortiz Nieto-Márquez et al. 2024; Masriera Esquerra and Santacana i Mestre 2012).

As mentioned earlier, **archaeology** is a science that combines multiple disciplines to formulate hypotheses about past societies through the remains found. Therefore, we will adopt the first step of archaeological methodology in site investigation, which is surveying.

Archaeological surveying is an active search process for potential signs indicating the presence of buried remains (such as buildings, roads, necropolises, walls…) (Ruiz-Zapatero 2004). It can be done aerially, using remote sensing techniques (Chivueco 1995; Rejas and Burillo 2016), such as aerial photography, or on the ground, either by walking to search for surface indicators or using geophysical techniques (Goodman and Piro 2013). None of these techniques are destructive, as they aim to preserve heritage (Malfitana et al. 2015; Sarris et al. 2013).

Aerial photography is a commonly used technique in archaeology to locate new sites, particularly because it saves time and effort. In this type of survey, differential growth marks in vegetation are very useful, as plant growth is influenced by the type of substrate (Rouse et al. 1973). Where there is a stone structure, plant roots cannot penetrate as deeply, resulting in reduced plant growth above these areas. However, if negative elements (ditches, pits, trenches…) are present beneath the plants, they will grow more because they are rooted in a less compact, more nutrient-rich substrate with higher organic content (Bennett 2012; Webber et al. 2019).

To obtain these images, we use geographic viewers like IBERPIX or Google Earth, whose timeline tool allows us to observe the same location during different periods (different phases of plant growth). The use of these ICT tools in the classroom supports competency-based learning and enables students to handle their own sources and data (Sebastián-López and De Miguel-González 2017).

On the other hand, for ground surveying, we have turned to **Geophysics**. Geophysics, as a discipline that studies the physical properties of the Earth, is applicable in this proposal as a tool to address the proposed challenge within the project.

The application of this discipline in classrooms has been successfully tested, particularly in higher education levels (Boyd and Romig 1997; Stewart et al. 2010). However, considering its potential, it is highly interesting to explore its integration into other educational levels.

One of the educational values we can derive from geophysics is that its techniques, applied in archaeological surveys, allow for non-invasive and non-destructive heritage studies, aligning with one of the project's objectives (Rivero and López Benito 2013; Vallés et al. 2023; Bellón Ruiz et al. 2021; Rascón Marqués et al. 2023). This approach ensures the conservation and care of heritage remains for their transmission to future generations.

Furthermore, it supports the idea of multi-interdisciplinarity in classrooms. To understand the operation of these technologies and their outcomes in a simplified manner, it is necessary to engage with various disciplines, including geology, physics, geography, mathematics, history, technology, and programming. Although no direct application has been carried out with primary school students, this proposal presents it as a challenge and an opportunity to broaden the range of adapted knowledge.

It is important to note that, depending on the educational level at which this proposal is applied, geophysical techniques may be somewhat complex. Therefore, we will introduce them by comparing them to something as familiar as certain medical tests (a common context for students).

A parallel can be drawn with the field of medicine, where techniques are used to view and interpret the interior of our bodies, such as X-rays or ultrasounds. These techniques rely on physical properties to create images and models that show bones or organs. In geology and archaeology, there are similar methods and techniques used to scan and document underground elements.

These techniques fall within the field of geophysics. Each material is different and possesses a set of chemical and physical characteristics that make it unique, like an ID. These characteristics and properties can be measured using high-precision and high-resolution equipment. Just like X-rays, images of the subsurface can be obtained, revealing the characteristics of each buried element and its position in depth, allowing us to know what lies beneath our feet. This is how petroleum, gold or silver, minerals and rocks, groundwater, archaeological structures, tombs, etc., can be located.

There are several data acquisition techniques based on different physical properties. In archaeology, methods are used that cover large areas in a short time and offer high resolution, allowing us to detect small objects (Goodman and Piro 2013). For these reasons, the most commonly used techniques are ground-penetrating radar (GPR), geomagnetism, and electromagnetism.

In our case, we will use *GPR*. It emits an electromagnetic pulse, a wave that penetrates the ground and receives the reflection or echo of the elements it encounters, producing graphics that show these reflections (Goodman et al. 2011; Zhao et al. 2013). This technique is widely used in non-invasive archaeological surveys because it is non-destructive, provides high-quality data (including depths of the remains), and is cost-effective.

Over the past few years, an increase in surveys using non-invasive techniques and their successful results can be documented (Bellón Ruiz et al. 2021; Colmenarejo García et al. 2023; Conyers 2015; Neubauer et al. 2002; Leucci et al. 2016; Piro and Campana 2012; Zhao et al. 2015, 2018).

Once the terrain has been analyzed and the results processed, they are visualized with **Geographic Information Systems (GIS).** Burrough and McDonnell (1998) defined GIS in 1998 as "a powerful set of tools for collecting, storing, retrieving, transforming, and displaying geospatial data from the real world for a set." However, most authors characterize GIS by its ability to model territory, its integrative potential for heterogeneous information sources, and its spatial analysis dimension, as well as the capability to perform geoprocessing operations based on the topological relationships of objects and the ability to generate new information and explain existing data (González et al. 2012).

The software chosen and used for this educational proposal is ESRI's ArcGIS, specifically ArcMap 10.8.1 and ArcScene 10.8.1. There is a wide variety of programs available, including free options such as QGIS, GRASS, and gvSIG. However, ArcGIS was selected because the university centre has a license, allowing us to work online with schools through ArcGIS Online.

Geographic Information Systems (GIS) are an educational tool with high potential for primary and secondary classrooms due to their ability to integrate multiple disciplines and foster active learning.

In primary education, GIS allows students to work with basic spatial concepts, promoting an understanding of their surroundings and the development of orientation and spatial analysis skills (Alibrandi and Goldstein 2015; Baker 2005).

In secondary education, their application expands to the analysis of complex geographic data, enabling students to explore topics such as climate change, natural resource distribution, or demographics, connecting geography with history, natural sciences, mathematics, and technology (Alibrandi and Goldstein 2015; Bednarz 2004; Milson et al. 2012).

Furthermore, GIS encourages critical thinking and problem-solving by addressing real-world issues through project-based learning and the use of innovative technological tools (Kerski 2008). This multidisciplinary nature makes GIS an ideal resource for developing digital, scientific, and civic competencies, preparing students to understand and tackle the challenges of today's world.

In terms of heritage, this tool is of utmost importance due to its capacity for managing and documenting heritage elements (Bru Castro et al. 2008; Farjas et al. 2009; Faustino Cordero and Álvarez 2017; Faustino Cordero Montero et al. 2020); its ability to create predictive models (Allen et al. 1990; Molina Martín 2009; Fernández-Fernández 2010, 2013; Fondevilla et al. 2021; Rascón Marqués et al. 2023); perform territorial analyses (Moreno Martín 2008; García Sanjuán et al. 2009; Ortiz et al. 2013; Ortiz Nieto-Márquez and Baena Preysler 2015; Ortiz Nieto-Márquez 2015), and even visibility and optimal route analyses (Zamora 2007, 2008, 2013; Orengo and Fiz 2008).

Additionally, GIS allows for the visualization of different data layers in an integrated way, with transparency, superimposed, or even modeled in 3D (Vallés-Iriso et al. 2019). This is the final solution for the dissemination, diffusion, and resources for Heritage Education in school classrooms, as well as for competency-based learning (Martín Vecino et al. 2016; Sánchez Cabielles 2014).

Practical Case

Based on the methodology and techniques described, a didactic proposal is suggested for application in a Primary/Secondary Education classroom, although it can always be adapted to higher or lower levels.

The Roman site of Caraca has been chosen as an example. This oppidum of the Carpetani tribe and Roman city is located at Cerro de la Virgen de la Muela (Driebes,

Guadalajara). The site was discovered in the twentieth century due to the discovery of silver remains on the hillside (Fernández and Gamo 2019). Thanks to subsequent prospecting campaigns and archaeological excavations, it has been confirmed that the hill has been occupied since the Paleolithic period, transitioning to a Carpetani oppidum upon which the Roman city was built. The city's peak occurred between the first and second centuries AD, and it began to decline and be abandoned in the third century AD (Gamo Pazos et al. 2023).

Caraca is a high imperial Roman city organized around a forum, cardo, and decumanus, with approximately 27 blocks and covering nearly 8 hectares (Fernández and Gamo 2019). The interest in this city lies in the high quality of the information obtained through prospecting processes and the potential it holds for classroom presentation.

The **didactic proposal** is divided into three sections. It begins with conflicting concepts (Excavation vs. Conservation), moves on to finding solutions to this problem through the application of new prospecting techniques, and finally concludes with an investigation into Heritage using non-invasive methods (Fig. 1).

In the **first introductory section**, Archaeology is introduced by asking students what they know about the discipline, what comes to mind when they think of it, and if they have ever visited an archaeological site. After this introduction, the teacher explains what Archaeology is, what can be documented at archaeological sites, and how this documentation is carried out.

Following this initial approach, students can be shown a news story about looting or the destruction of archaeological remains, prompting them to reflect on whether such actions are right or wrong. In this context, it would be explained that everything

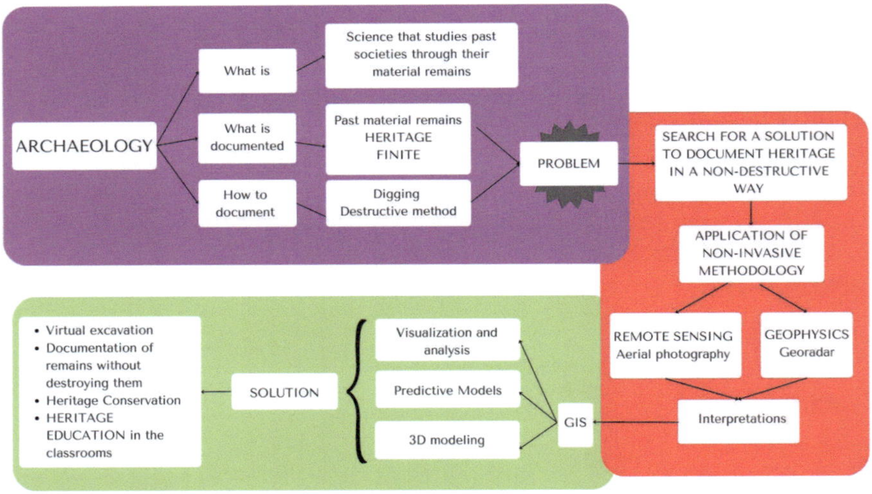

Fig. 1 Schematic representation of the educational proposal based on PBL for Heritage Education. *Source* Own elaboration

Fig. 2 Schematic representation of the problem as the central axis of the research process. *Source* Own elaboration

documented at archaeological sites is part of our cultural heritage, and as citizens, we must conserve and preserve it.

This is where students should become aware of the contradiction between the destructive methods of archaeological excavations and the obligation to preserve heritage as outlined by UNESCO (1972). This problem will be presented as the central focus of the project, prompting them to reflect, analyze the situation, and seek non-destructive methods to document archaeological sites (Fig. 2).

The **second block** will be organized around this real problem, where students must seek alternatives and create their hypotheses. They will be able to research in the classroom and present ideas as possible solutions. Later, examples of aerial prospections can be shown, featuring different buried but visible sites due to the differential growth of vegetation (Fig. 3). Based on these images, students can work directly with geographic viewers and infer what might be underground.

Similarly, examples of results from the interpretations of geophysical prospections of certain sites can be shown next, prompting students to think about what they are seeing and how the interpretations were made (Fig. 4). After this reflective process, geophysics will be introduced, explaining what it is and its applications, one of which is archaeology.

With these two techniques, archaeological data can be obtained without the need for excavation, thus preserving the heritage.

Finally, we begin the **third block.** To visualize and interact with the results of the prospections, Geographic Information Systems (GIS) will be used. Aerial photographs can be overlaid with georadar interpretations (Fig. 5). This step allows for the comparison of the prospection techniques and determines which one yields better results or if they can be used in conjunction.

Once the results of the ground and aerial prospections are georeferenced, a map of the interpretations is obtained. This map displays the urban layout of the city, which adapts to the morphology of the hill. At this stage, students can visualize what the city would look like without the need to excavate or even visit the site (Fig. 6).

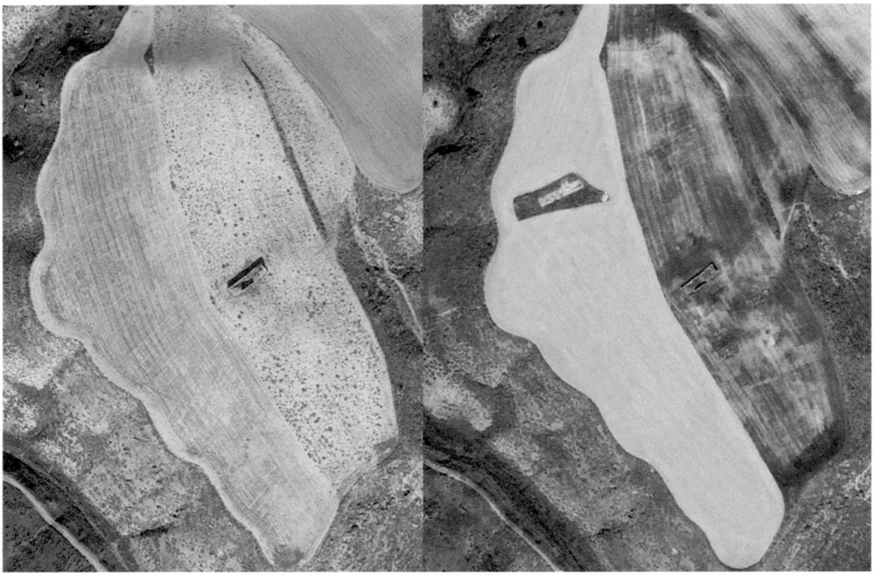

Fig. 3 Aerial photograph of the hill where the Caraca archaeological site is located (Driebes, Guadalajara). Images taken from Google Earth; the one on the left is from 2015, and the one on the right is from 2024. Differential growth of vegetation can be observed. *Source* Own elaboration

They can document the most important parts of the city, such as the forum, the baths, the circulation routes, large blocks that could represent the villas of wealthy families, irregular layouts, and smaller structures to the south…

Thanks to the various tools provided by GIS, it is possible to answer some questions that may arise after visualizing the results. In this case, the absence/presence of remains in certain areas of the hill is striking, leading to the following question: What could have happened? (Fig. 7).

By examining the GPR alteration map (Fig. 6), we can identify areas with greater clarity and depth, as well as areas with fainter anomalies and lesser thickness, alongside zones with a complete absence of anomalies, where continuity should exist.

The Digital Terrain Model (DTM) of the hill shows that in the highest areas, structures are thinner, and the soil is more heterogeneous, indicating that these regions have experienced greater abrasion due to human actions or higher erosion due to their elevation. Conversely, in the peripheral areas of this elevated zone, the structures are more distinct (showing greater contrast between the structures and their surroundings, which suggests a more homogeneous soil), indicating better preservation.

Through hydrological analysis, the heads of the small streams or gullies on the hill have been located, which have been altered by current plowing practices that tend to flatten cultivated surfaces. It can be seen that these heads coincide with the areas of absence of remains on the anomaly map, suggesting that the site has been

Fig. 4 Example of results from geophysical surveys with ground-penetrating radar in archaeological sites. Software: ArcMap 10.8.1. *Source* Own elaboration

affected, potentially disappearing due to erosive actions before agricultural use of the hill (Fig. 7).

These issues address post-depositional processes that directly affect our heritage. The fact that archaeological sites are exposed to the elements, as is the case with most sites, means that weather phenomena and human activities impact the condition and conservation of these assets.

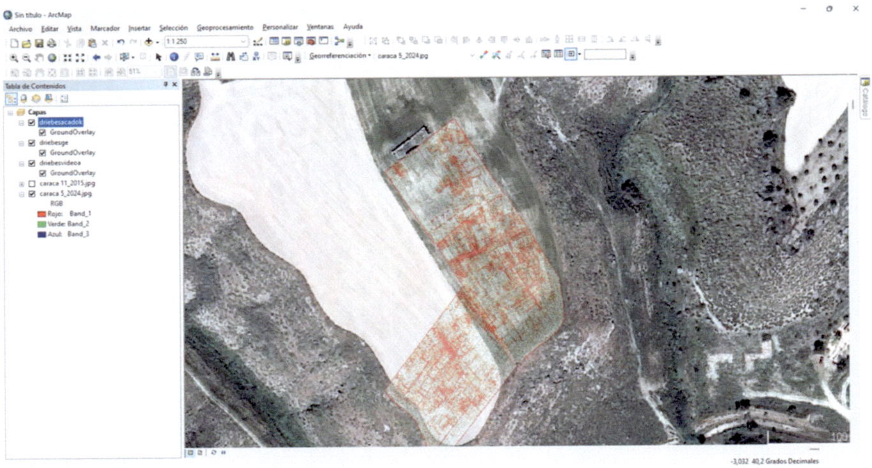

Fig. 5 ArcGIS interface with the overlay of the photograph showing differential vegetation growth and the results of the ground-penetrating radar. Software: ArcMap 10.8.1. *Source* Own elaboration

Fig. 6 Comparison of the results obtained from the GPR and their interpretation. Software: ArcMap 10.8.1. *Source* Own elaboration

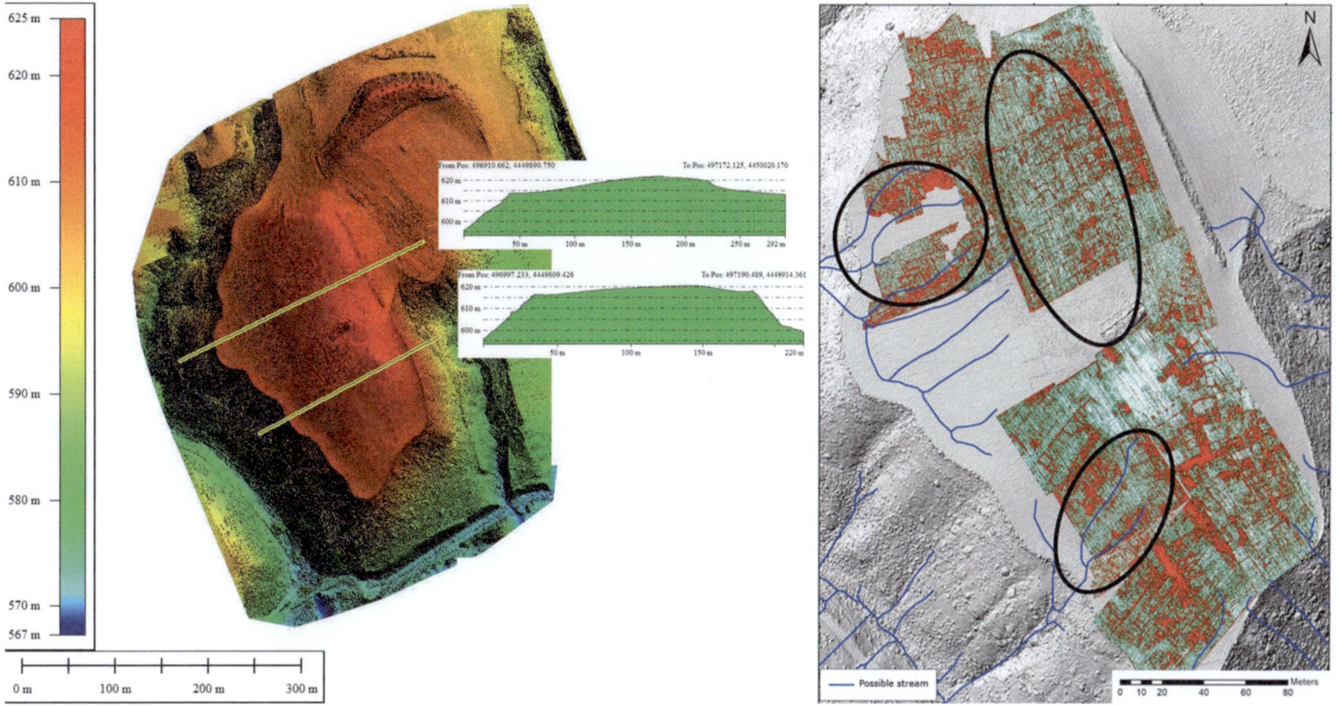

Fig. 7 Use of the Digital Terrain Model (DTM) of the hill to create cross-sections and observe its topography, allowing for comparison with the absence/presence of remains, along with the influence of hydrological processes. Software: ArcMap 10.8.1. *Source* Own elaboration

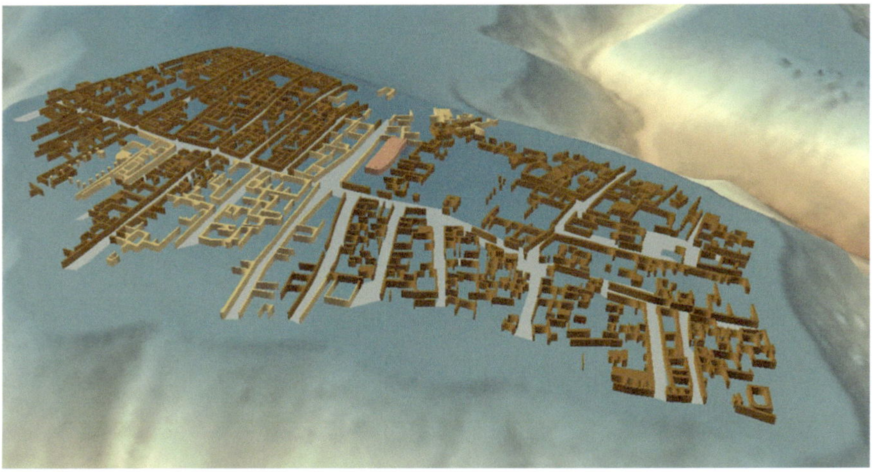

Fig. 8 3D modeling of the interpretations of non-invasive prospections. Software: ArcScene 10.8.1. *Source* Own elaboration

In this instance, the site was affected by water runoff caused by rainfall, which led to the deterioration of certain areas of the urban landscape. Subsequently, with the change in land use to agricultural, farming activities have also impacted its preservation. Considering this dilemma between socio-economic advancement and heritage conservation, a debate can be initiated in the classroom.

Finally, thanks to ArcGIS's capability to work with 3D elements in ArcScene, a model can be created using the DTM of the site as a base for heights, onto which the interpretations from the surveys are overlaid, with extrusions to reach the desired height (Fig. 8).

It is with this tool that students get a more realistic view of what the city of Caraca could have been. This is the final moment of generalization, where the final solutions (aerial photography, use of geophysics, etc.) are proposed, and students can evaluate the process they followed.

Conclusions

When conflicting cases arise in the classroom, such as the current one where documenting heritage remains would require excavating and destroying them, while legislation advocates for conservation, students tend to reflect on the dilemma, presenting them with a real problem they must confront.

This is one of the primary objectives of Heritage Education (educating from, by, and for heritage): to make society part of the heritage, to feel it as their own, and at the same time to understand that they can generate it for future generations. By analyzing the issues presented in the classroom, students must become aware of the importance of the remains, as they form part of cultural heritage and history. When certain heritage elements are controversial or generate debate, they can contribute the most to Heritage Education. These actions ensure that students actively participate in their learning through their critical and reflective abilities (Cuenca López 2023).

They must analyze the situation, determine whether it is correct or not based on social values, and take action. This is why heritage education in classrooms is so important. In this case, it has been linked to archaeology and non-invasive techniques, but many other disciplines can be addressed if we analyze the conservation and restoration of heritage elements, or if we deal with the conflict or integration of leisure and tourism related to heritage.

Pedagogically, this problem-based learning (PBL) proposal fits perfectly with the treatment of heritage education in the classroom, especially focused on controversial heritage. One should not fall into the temptation of viewing the problem as Heritage itself; rather, the issue lies in the damage and destruction it suffers, which must be avoided.

This learning approach, which combines PBL and Heritage Education, stimulates students' critical thinking as they investigate to propose solutions. Its application allows for effective evaluation of the knowledge acquired. Additionally, it promotes cognitive development by encouraging comparison and exchange of ideas with others, and fosters active participation in problem-solving (Cuenca López 2013; Eggen and Kauchak 2015; Estepa Jiménez and Cuenca López 2006; Monfort 2007).

The use of new technologies in the classroom, such as GIS, cartographic viewers, and the presentation of data obtained through new non-invasive prospecting techniques, reinforces the role of ICT in the learning processes (LOMLOE), while strengthening the students' relationship with heritage.

GIS plays a crucial role in acquiring geographic, historical, artistic, and heritage competencies and knowledge.

As can be seen in the Table 1, the proposed approach is fully immersed in competency acquisition. Key competencies, as well as specific ones, are achieved whether implemented in Primary or Secondary Education (LOMLOE).

Finally, it can be concluded that the proposed approach combines new technologies (cartographic viewers and GIS) with disciplines such as archaeology, geophysics, and remote sensing to serve heritage, its conservation, and its dissemination and promotion in society (in this case, the classrooms). This work demonstrates how Heritage Education can be conducted in classrooms in an active, participatory, and immersive manner, where students are part of their learning by seeking solutions to a real problem that affects the conservation and legacy of the remains (archaeological in this case).

Table 1 Competency application of the applied disciplines in the proposal

DISCIPLINES	KEY COMPETENCIES	Operational Descriptor
HERITAGE EDUCATION	Citizenship Competence	CC1, CC2 & CC4
	Cultural Awareness and Expression Competence	CCEC1, CCEC2 & CCEC 4
ARCHAEOLOGY	Digital Competence	CD1, CD2, CD3 & CD5
	Cultural Awareness and Expression Competence	CCEC1 & CCEC2
	Citizenship Competence	CC1, CC2, CC3 & CC4
	Mathematical and Scientific Competence in Engineering	STEM2, STEM4 & STEM5
GEOPHYSICS	Mathematical and Scientific Competence in Engineering	STEM2, STEM4 & STEM5
	Digital Competence	CD1, CD2, CD3 & CD5
	Digital Competence	CD1, CD2, CD3 & CD5
	Entrepreneurial Competence	CE1
	Mathematical and Scientific Competence in Engineering	STEM2, STEM4 & STEM5
PRIMARY SPECIFIC COMPETENCIES	SECONDARY SPECIFIC COMPETENCIES	
Knowledge of the Natural, Social, and Cultural Environment: CE1, CE2, CE3, CE5, CE6 & CE7	Geography and History: CE1, CE2, CE3, CE4, CE5 & CE7	

Source Own elaboration

Acknowledgements We would like to thank the research team at the Caraca site (Driebes, Guadalajara) for their generous collaboration. Their contribution has been essential for the development and enrichment of this work.

Annex

INSTRUCTIONAL DESIGN FOR PBL		
	PRIMARY EDUCATION (5th and 6th grades - ages 10–12)	**SECONDARY EDUCATION** (3rd and 4th grades of ESO - ages 14–16)
1. PLANNING **LEARNING OBJECTIVES**	1. Understand what archaeological heritage is and its importance. 2. Become familiar with basic concepts of ground-penetrating radar (GPR) and remote sensing. 3. Develop observation and data recording skills. 4. Use simple ICT tools to interpret maps.	1. Analyze non-invasive documentation technologies (GPR, remote sensing). 2. Interpret data obtained using Geographic Information System (GIS) tools. 3. Propose solutions for the conservation of archaeological heritage based on findings. 4. Reflect on the importance of heritage conservation for cultural identity.
1. PLANNING **KEY COMPETENCES**	1. Competence in cultural awareness and expression. 2. Digital competence. 3. Mathematical competence and basic competences in science and technology. 4. Competence in learning to learn. 5. Social and civic competences.	1. Competence in cultural awareness and expression. 2. Digital competence. 3. Mathematical competence and basic competences in science and technology. 4. Competence in linguistic communication. 5. Competence in learning to learn. 6. Social and civic competences.
1. PLANNING **TRANSVERSAL COMPETENCES**	- Teamwork and collaboration - Problem-solving - Environmental awareness - Critical thinking	
1. PLANNING **PROBLEM**	How can we discover and protect an archaeological site without damaging it? A team of archaeologists has found evidence of an important archaeological site near your community. However, they cannot excavate directly because the site is protected by environmental laws and is located in a nature reserve. They need your help to document and study the site using non-invasive methods such as ground-penetrating radar (GPR) and remote sensing. Additionally, you will propose solutions to protect the site and share it with the public without causing harm.	
1. PLANNING	6 SESSIONS OF 1 HOUR	
2. GROUP ORGANIZATION	25 students: 5 groups of 5 students	30 students: 6 groups of 5 students
3. BRAINSTORMING	Students share their knowledge about Archaeology. There is no debate, and everything is noted down.	
4. PROBLEM PRESENTATION	The problem is presented, and terms are clarified. Present the case to the students and address any questions. Specify the time available and the evaluation criteria.	
5. PROBLEM DEFINITION	Students analyze the case and identify the problem. They must express it in a single question or statement	
6. RESPONSE AND HYPOTHESIS FORMULATION	Students must explore alternatives and create their hypotheses. They can also conduct research in the classroom and present ideas as possible solutions.	

7. FORMULATION OF LEARNING OBJECTIVES	**Cognitive Objectives** -Understand what archaeological heritage is and why it is important to protect it. -Learn what non-invasive technological tools (ground-penetrating radar, remote sensing) are and how they work in a basic and engaging way. -Identify basic cultural and historical elements in the context of the fictional site (structures, objects). **Procedural Objectives** -Use a simple ground-penetrating radar simulator to detect patterns or shapes in the terrain. -Record data in a field journal, organizing the information clearly. -Create an illustrated map of the fictional site with their findings and explain its significance to a non-specialized audience. **Attitudinal Objectives** -Foster curiosity and interest in the conservation of cultural heritage. -Collaborate in teams, respecting the ideas of peers. -Appreciate the importance of preserving the past for future generations	**Cognitive Objectives** -Analyze in depth what archaeological heritage is and how it relates to cultural identity. -Understand the technical functioning of tools like ground-penetrating radar (GPR) and remote sensing systems. -Introduce basic concepts for interpreting geospatial data using Geographic Information Systems (GIS). -Recognize the environmental, social, and cultural risks that threaten heritage conservation. **Procedural Objectives** -Use GIS software to load, analyze, and represent data obtained through non-invasive technology. -Design conservation proposals for the site, considering social and environmental factors. -Prepare and present a well-argued report that integrates historical, technological, and strategic analyses. **Attitudinal Objectives** -Develop a critical attitude towards the relationship between technology, sustainability, and heritage. -Appreciate interdisciplinary work and the importance of collaboration in complex projects. -Reflect on individual and collective responsibility in the protection of cultural heritage.
8. RESEARCH	To visualize and interact with the results of the surveys, GIS tools will be used.	
9. SYNTHESIS AND PRESENTATION	Synthesis: Creative and visual (illustrated maps, posters). Presentation Level: Playful and narrative to engage the audience. Simulated Audience: School or family community.	Synthesis: Analytical and technical (digital maps, reports). Presentation Level: Formal presentation with reasoning and concrete data. Simulated Audience: Technical or institutional committee.
10. EVALUATION AND SELF-ASSESSMENT	**Cognitive Evaluation** -Questionnaires or Basic Knowledge Rubrics (self-assessment and peer assessment). **Procedural Evaluation** -Direct observation. -Visual Production Rubric. -Simple field report. **Attitudinal Evaluation** -Rubric for collaboration and group attitude (peer assessment). -Self-assessment.	**Cognitive Evaluation** -Short test. -Open-ended questions. **Procedural Evaluation** -Rubric for the technical report and digital presentation. -Evaluation of the technical quality of the GIS map. -Presentation of the conservation proposal. **Attitudinal Evaluation** -Group evaluation rubric. -Group self-assessment.

References

Alibrandi M, Goldstein D (2015) Integrating GIS and other geospatial technologies in middle schools. In: Geospatial technology and geography education in a changing world: a critical reader. Springer, pp 91–104. https://doi.org/10.1007/978-4-431-55519-3_5

Allen K, Green S, Zubrow E (eds) (1990) Interpreting space: GIS and archaeology. Taylor and Francis, London

Ayán Vila XM, Criado Boado F, González Veiga M, Otero Vilariño C (2009) Cultura científica en arqueología y patrimonio: Los valores educativos de lo invisible. En: *V Congreso Internacional Musealización de Yacimientos Arqueológicos, Cartagena 2008*, pp 115–124. ISBN 978-84-96728-18-9

Baker TR (2005) Internet-based GIS mapping in support of K-12 education. Prof Geogr 57(1):44–50

Bednarz SW (2004) Geographic information systems: a tool to support geography and environmental education? GeoJournal 60(2):191–199

Bellón Ruiz JP, Lechuga Chica MÁ, Rueda Galán C, Moreno Padilla MI, Quesada Sanz F, Molinos Molinos M, Ruiz Rodríguez A, García-Bellido MP, Nieto-Márquez IO, Vallés Iriso J (2021) De situ Iliturgi, análisis arqueológico de su asedio en el contexto de la segunda guerra púnica. Archivo Español de Arqueología 94:e15. https://doi.org/10.3989/aespa.094.021.15

Bennett R et al (2012) The application of vegetation indices for the prospection of archaeological features in grass-dominated environments. Archaeol Prospect 19(3):209–218. https://doi.org/10.1002/arp.1429

Boyd TM, Romig PR (1997) Cross-disciplinary education: the use of interactive case studies to teach geophysical exploration. Comput Geosci 23(5):593–599. https://doi.org/10.1016/S0098-3004(97)00028-9

Bru Castro MA, López Quiroga J, Farjas Abadía M, Zancajo Jimeno JJ, Martínez Tejera AM (2008) Cultural resource management medieval team. Projects and applications in the Iberian Peninsula. In: On the road to reconstructing the past, computer applications and quantitative methods in archaeology conference, program and abstracts CAA 2008. Archaeolingua Foundation, Budapest, pp 155–156

Burrough PA, McDonnell RA (1998) Principles of geographical information system. Oxford University Press, USA

Castillo Lozano JÁ (2017) La importancia de la arqueología en las aulas: La recreación de una excavación arqueológica como actividad innovadora en la educación del alumnado de la E.S.O. Revista Otarq 2:411–428. https://doi.org/10.23914/otarq.v0i2.125

Chivueco E (1995) Fundamentos de la teledetección espacial. Ediciones RIALP S.A.

Colmenarejo García F, de Andres Santos M, Vallés Iriso J, Ortiz Nieto-Márquez I, Herranz Rodrigo D (2023) Historiografía y nuevas perspectivas de la Necrópolis de Fuente del Moro (Colmenar Viejo), a partir de las prospecciones geofísicas mediante georradar 3D. Estudios de Historia y Patrimonio 1:17–43

Conyers LB (2015) Analysis and interpretation of GPR datasets for integrated archaeological mapping. Near Surf Geophys 13(6):645–651. https://doi.org/10.3997/1873-0604.2015018

Cordero F, Álvarez LA (2017) APP CÁCERES VIEW. La IDE local al servicio del usuario. Revista Mapping 26:20–26. http://ide.caceres.es

Cordero Montero F et al (2020) Arqueología del planeamiento urbano de Cáceres. Revista Mapping 29:22–28

Cuenca López JM (2013) El papel del patrimonio en los centros educativos: hacia la socialización patrimonial. Tejuelo 19:76–96

Cuenca López JM (2023) Preliminares: Los patrimonios desde la perspectiva del conflicto. Los problemas controversiales a través de la educación patrimonial. REIDICS 12:6–12. https://doi.org/10.17398/2531-0968.12.01

Dietz C, Catanzariti G, de la Presa P, Jimeno A (2011) Técnicas no invasivas para el registro del patrimonio arqueológico. Jornadas Sobre Documentación Gráfica Del Patrimonio Histórico. Presente y Futuro, pp 118–129

Egea A, Arias L (2013) IES Arqueológico: La arqueología como recurso para trabajar las competencias básicas en la educación secundaria. Clío: History and History Teaching, 39
Eggen P, Kauchak D (2015) Estrategias docentes. Enseñanza de contenidos curriculares y desarrollo de habilidades de pensamiento. Fondo de Cultura Económica, México
Estepa Jiménez J, Cuenca López JM (2006) La mirada de los maestros, profesores y gestores del Patrimonio. Investigación sobre concepciones acerca del Patrimonio y su didáctica. In Calaf R, Fontal O (eds) Miradas al Patrimonio. TREA, pp 51–72
Faltýnová M et al (2015) Complex archaeological prospection using combination of non-destructive techniques. Int Arch Photogramm Remote Sens Spat Inf Sci XL-5/W7:141–146. https://doi.org/10.5194/isprsarchives-XL-5-W7-141-2015
Farjas M, García-Lázaro FJ, Zancajo J, Mostaza T (2009) Automatic point-cloud Surveys in prehistoric site documentation and modelling. In: Fischer L, Frischer B, Wells S (eds) Making history interactive, 37th computer applications and quantitative methods in archaeology CAA 2009, online proceedings. Williamsburg, Virginia. http://www.caa2009.org/articles/Farjas_Contribution163_a.pdf
Faustino Cordero Montero LA, Álvarez Llorente AJ, Gómez González AJ, Sánchez Franco C (2020) Arqueología del planeamiento urbano de Cáceres. Revista Mapping 29:22–28. ISSN: 1131-9100
Faustino Cordero LA (2017) APP CÁCERES VIEW. La IDE local al servicio del usuario. Revista Mapping 26:20–26. ISSN: 1131-9100
Fernández Fernández J (2010) Modelos predictivos y SIG: algunos ejemplos de su aplicación en Arqueología: los asentamientos fortificados del Valle del Trubia (Asturias, España). Mapp Sci Remote Sens
Fernández Fernández J (2013) 'Arqueología, SIG y modelos predictivos basados en redes neuronales artificiales Un caso de Estudio. Mapping 160:30–36
Fernández J, Gamo E (2019) Hallazgo de un tesoro de plata en época de Felipe II en Driebes (Guadalajara). In Gamo E, Fernández J, Álvarez D (eds) En ningún lugar... Caraca y la romanización de la Hispania interior. Guadalajara, pp 189–218.
Fondevilla JJ, Rivera Jiménez T, Ampliato Briones AL (2021) Implementación SIG de modelos analíticos predictivos del sistema castramental dispuesto por el concejo de Sevilla en torno a la Vía de la Plata en la Baja Edad Media. Arqueología De La Arquitectura 18:e111. https://doi.org/10.3989/arq.arqt.2021.003
Fontal Merillas O (2006) Introducción. El Patrimonio mirado desde diferentes ópticas. In Calaf R, Fontal O (eds) Miradas al Patrimonio, Ediciones Trea, pp 17–19
Gamo Pazos E, Fernández Ortea J, Bustamante Álvarez M, Marín González S, Domínguez Solera SD, González Blas A (2023) Un Ustrinum romano en Caraca (Driebes, Guadalajara). ANAS 36:259–284
García Sanjuán L, Wheatley D, Márquez Pérez J (2009) Los SIG y el análisis espacial en arqueología: aplicaciones en la prehistoria reciente del sur de España. In Cau Ontiveros MA, Nieto Prieto FX (eds) Arqueología náutica mediterránea. Centre d'Arqueologia Subacuàtica de Catalunya, Girona, pp 163–180
González Marcén P (2010) La dimensión educativa de la arqueología. In: La tutela del patrimonio prehistórico. Congreso Memorial Siret. Antequera. Disponible en: http://bit.ly/1nK8jLS
González IDB, Freire CF, Morente LM-F, Asensio EP (2012) Los Sistemas de Información Geográfica y la investigación en ciencias humanas y sociales. In: Apuntes de Ciencias Instrumentales y Técnicas de Investigación, 3. CCHS-CSIC, Madrid
Goodman D, Novo A, Morelli G, Piro S, Kutrubes D, Lorenzo H (2011) Advances in GPR imaging with multi-channel radar systems from engineering to archaeology. SAGEEP, 416–422. https://doi.org/10.4133/1.3614128
Goodman D, Piro S (2013) GPR remote sensing in archaeology. SPRINGER-VERLAG BERLIN AN, Geotechnology and Enviroment
Henson D (2004) Archaeology and education, an exercise in constructing the past. Treballs d'Arqueologia 10:5–16

Kerski JJ (2008) The role of GIS in digital earth education. Int J Digit Earth 1(4):326–346
Leucci G, de Giorgi L, di Giacomo G, Ditaranto I, Miccoli I, Scardozzi G (2016) 3D GPR survey for the archaeological characterization of the ancient Messapian necropolis in Lecce, South Italy. J Archaeol Sci Rep 7:290–302. https://doi.org/10.1016/j.jasrep.2016.05.027
Malfitana D et al (2015) The potential of integrated GPR survey and aerial photographic analysis of historic urban areas: a case study and digital reconstruction of a Late Roman villa in Durrës (Albania). J Archaeol Sci Rep 4:276–284. https://doi.org/10.1016/j.jasrep.2015.09.018
Marra R, Jonassen DH, Palmer B, Luft S (2014) Why problem-based learning works: theoretical foundations. J Excel CollE Teach 25(3–4):221–238
Martín Vecino JM, Nieto Masot A, Buzo Isaac (2016) Los SIG aplicados a la enseñanza de la geografía en 1° de educación secundaria obligatoria. In: TECNOLOGÍAS DE LA INFORMACIÓN GEOGRÁFICA EN EL ANÁLISIS ESPACIAL. Aplicaciones en los Sectores Público, Empresarial y Universitario, pp. 141–160. Grupo de Investigación en Desarrollo Sostenible y Planificación Territorial de la Universidad de Extremadura Grupo de Investigación Geo-Ambiental de la Universidad de Extremadura. https://www.researchgate.net/publication/312054015
Masriera Esquerra C, Santacana i Mestre, J (2012) La arqueología reconstructiva y el factor didáctico. Biblioteconomía y Administración Cultural - 241, 1 ed., Gijon
Micó Pérez R, Benito López JE, Ruiz Zapatero G, Rodríguez Ariza MO, Cotino Villa F, Burillo Mozota F, Molinos Molinos M, Picazo Millán JV (1998) Metodología para la recuperación del dato arqueológico. Arqueología espacial 18:45–56
Milson AJ, Demirci A, Kerski JJ (eds) (2012) International perspectives on teaching and learning with GIS in secondary schools. Springer. https://doi.org/10.1007/978-94-007-2120-3
Molina Martín O (2009) Aplicación de los SIG para el desarrollo de modelos predictivos en la prospección y búsqueda de contextos arqueológicos. Boletín Antropológico 27(76):219–236
Monfort NG (2007) El valor educativo y uso didáctico del Patrimonio Cultural. Universitat Autónoma de Barcelona, España
Montoya Martínez FJ, Egea Vivancos A (2021) La arqueología experimental como estrategia educativa: realidad y posibilidades. Investigación en la Escuela 103:139–152. https://doi.org/10.12795/IE.2021.i103.10
Moreno Martín A (2008) Using GIS to Deconstruct Iberian Iron Age Landscapes: The Territory of Kelin Between 6th–5th Centuries BC (La Plana d'Utiel, València). In: Huggett J (ed) Layers of perception. Proceedings of the 35th computer applications and quantitative methods in archaeology CAA 2007. Archaeopress, Oxford, pp 87–92
Neubauer W, Eder-Hinterleitner A, Seren S, Melichar P (2002) Georadar in the Roman civil town Carnuntum, Austria: an approach for archaeological interpretation of GPR data. Archaeol Prospect 9(3):135–156. https://doi.org/10.1002/arp.183
Orengo HA, Fiz I (2008) The application of 3D reconstruction techniques in the analysis of ancient Tarraco's urban topography. In Posluschny A, Lambers K, Herzog I (eds) Layers of perception. Proceedings of the 35th international conference on computer applications and quantitative methods in archaeology (CAA), Berlin, Germany, April 2–6, 2007, Kolloquien zur Vor- und Frühgeschichte, vol. 10. Dr. Rudolf Habelt GmbH, Bonn
Ortiz Nieto-Márquez I, Baena Preysler J (2015) ¿Acciones antrópicas o naturales en los hogares neandertales? Arqueología experimental y análisis intra-site. In: Maximiano Castillejo AM, Cerrillo Cuenca E (eds) Arqueología y tecnologías de información espacial: una perspectiva ibero-americana, pp 168–190. ISBN 978-1-78491-318-2
Ortiz Nieto-Márquez I, Baena Preysler J, Chacón MG (2013) GIS spatial distribution analysis in raw material quarrying sites: the example of El Cañaveral (Madrid, Spain). In: García Moreno A, García Sánchez J, Maximiano Castillejo AM (eds) Debating spatial archaeology: proceedings of the international workshop on landscape and spatial analysis in archaeology, Santander, June 8–9, 2012, pp 199–211. ISBN 978-84-697-0213-0
Ortiz Nieto-Márquez I, Muñoz García G, Jiménez Pablo E (2024) Arqueología en las aulas de Educación Infantil para maestros/as en formación: propuestas y recursos didácticos a partir de

La Casa de Hippolytus (Complutum, Alcalá de Henares). Revista UNES. Universidad, Escuela Y Sociedad 18:161–179. https://doi.org/10.30827/unes.i18.31163

Ortiz Nieto-Márquez I (2015) Acercamiento a las conductas neandertales a través del análisis espacial: Remontajes y SIG. Universidad Autónoma de Madrid, Tesis Doctoral

Piro S, Campana S (2012) GPR investigation in different archaeological sites in Tuscany (Italy): analysis and comparison of the obtained results. Near Surf Geophys 10(1):47–56. https://doi.org/10.3997/1873-0604.2011047

Rascón Marqués S, Vallés Iriso J, Sánchez Montes AL, Ortiz Nieto-Márquez I, Herranz Rodrigo D (2023) Nueva información sobre el urbanismo de la ciudad romana de "Complutum" a partir de los últimos estudios geofísicos: las prospecciones de 2020 y 2021. Anas 36:145–165

Rejas JG, Burillo F (2016) Teledetección aplicada a la arqueología. In Mínguez García MC, Capdevila Montes E (eds) Manual de tecnologías de la información geográfica aplicadas a la arqueología, pp 241–270. Comunidad de Madrid: Museo Arqueológico Regional ; Ilustre Colegio Oficial de Doctores y Licenciados en Filosofía y Letras y en Ciencias de la Comunidad de Madrid

Renfrew C, Bahn PG (1993) Arqueología: teorías, métodos y práctica. Ediciones Akal

Rivero V, López Benito V (2013) Virtual archaeology as an instructional tool. In: Virtual archaeology: nondestructive methods of prospections, modeling, reconstructions. Proceedings of the first international conference (The State Hermitage Museum 4–6 June 2012). San Petersburgo, pp 295–302

Rouse JW, Haas RH, Schell JA, Deering DW (1973) Monitoring vegetation systems in the great plains with ERTS. In: Third ETRS symposium, pp 309–317

Ruiz Zapatero G (2013) La excavación arqueológica. In: García Diez M, Zapata Peña L (Coords) Métodos y técnicas de análisis y estudio en arqueología prehistórica: De lo técnico a la reconstrucción de los grupos humanos. Universidad del País Vasco, pp 39–72

Ruiz Zapatero G (2010) Los valores educativos de la prehistoria en la enseñanza obligatoria. MARQ, Arqueología y Museos 04:161–179

Ruiz-Zapatero G (2004) La prospección arqueológica en los inicios del siglo XXI. Arqueología Espacial 24–25:17–32

Sánchez Cabielles P (2014) TIC y didáctica de la Geografía: el papel del SIG en la educación secundaria. Universidad de Cantabria, Tesis de Máster

Sarris A et al (2013) Integration of geophysical surveys, ground hyperspectral measurements, aerial and satellite imagery for archaeological prospection of prehistoric sites: the case study of Vészto-Mágor Tell, Hungary. J Archaeol Sci 40(3):1454–1470. https://doi.org/10.1016/j.jas.2012.11.001

Sebastián-López M, de Miguel-González R (2017) Educación Geográfica 2020: IBERPIX y Collector for ArcGIS como recursos didácticos para el aprendizaje del espacio. Didáctica Geográfica 18:231–246

Stewart RR, Khan S, Hall S, Liner C, Wong J (2010) Geophysics field education: better learning by doing. Lead Edge 29:546–550. https://doi.org/10.1190/tle29050546.1

Vallés J, Ortiz Nieto-Márquez I, Herranz-Rodrigo D, Yravedra J, Martínez-Pagán P, Segura MA, Bellón Ruiz JP, Lechuga MÁ, Díaz A, 2023. Más allá de los muros: Estructuras negativas y "falsos positivos". La alianza entre la arqueología y la geofísica. En: *10.ª Asamblea Hispano-Portuguesa de Geodesia y Geofísica*, pp 1070–1080. Catálogo de publicaciones de la Administración General del Estado. Disponible en: https://cpage.mpr.gob.es

Vallés-Iriso J, Ortiz Nieto-Márquez I, Chapa-Brunet MT, Barderas Manchado G, Yravedra Sainz de los Terreros J, Turégano-Botija M (2019) Una mirada al subsuelo: estudio del yacimiento romano del Cerro de la Virgen de la Muela (Driebes, Guadalajara) mediante georradar multicanal. In Gamo Pazos E, Fernández Ortea J, Álvarez Jiménez D (eds) En ningún lugar… Caraca y la romanización de la Hispania interior. Diputación provincial de Guadalajara, Guadalajara, pp 237–250

Vicent N, Rivero Gracia MP, Feliu Torruella M (2015) Arqueología y tecnologías digitales en Educación Patrimonial. Educatio Siglo XXI 33(1):83–102. https://doi.org/10.6018/j/222511

Webber H, Heyd V, Horton M, Bell M, Matthews W, Chadburn A (2019) Precision farming and archaeology. Archaeol Anthropol Sci 11(2):727–734. https://doi.org/10.1007/s12520-017-0564-8

Zamora M (2007) Total and cumulative viewshed: an application in the Genil River Valley. In: Figueiredo A, Velho G (eds) The world is in your eyes: proceedings of the XXXIII computer applications and quantitative methods in archaeology conference, CAA 2005. Tomar, Portugal, pp 313–318

Zamora M (2008) Thematic cartography in archaeology. In: On the road to reconstructing the past: computer applications and quantitative methods in archaeology conference, program and abstracts CAA 2008. Budapest, Hungary, pp 302–303. Archaeolingua Foundation

Zamora M (2013) Análisis territorial en arqueología: percepción visual y accesibilidad del entorno. Comechingonia: Revista de Arqueología 17(2):83–106. ISSN-e 2250-7728. ISSN 0326-7911

Zhao W, Forte E, Pipan M, Tian G (2013) Ground penetrating radar (GPR) attribute analysis for archaeological prospection. J Appl Geophys 97:107–117. https://doi.org/10.1016/j.jappgeo.2013.04.010

Zhao W, Forte E, Levi ST, Pipan M, Tian G (2015) Improved high-resolution GPR imaging and characterization of prehistoric archaeological features by means of attribute analysis. J Archaeol Sci 54:77–85. https://doi.org/10.1016/j.jas.2014.11.033

Zhao W, Forte E, Fontana F, Pipan M, Tian G (2017) GPR imaging and characterization of ancient Roman ruins in the Aquileia Archaeological Park, NE Italy. Meas J Int Meas Confed 113:161–171. https://doi.org/10.1016/j.measurement.2017.09.004

Legislation/Regulations

LOMLOE (2020) Ley Orgánica 3/2020, de 29 de diciembre, por la que se modifica la Ley Orgánica 2/2006, de 3 de mayo, de Educación. Boletín Oficial del Estado, núm. 340, 30 de diciembre de 2020, pp 122868–122953. https://www.boe.es/boe/dias/2020/12/30/pdfs/BOE-A-2020-17264.pdf

UNESCO (1972) Convención sobre la Protección del Patrimonio Mundial, Cultural y Natural. París, del 17 al 21 de octubre de 1972. Disponible en: https://whc.unesco.org/archive/convention-es.pdf

Ancient Labyrinths of Europe: New Opportunities for Systemic Research

A. Paranina and R. Paranin

Nature speaks to us in the language of mathematics...
Its letters are simple geometric figures: circles, triangles, lines...
without them, futile wanderings in the labyrinth... (Galileo Galilei 1623)

Abstract This chapter shows the possibilities of rational (instrumental) interpretation of the primary functions of prehistoric cultural heritage sites using the example of stone labyrinths located on the shores of the White Sea in the Republic of Karelia. It is shown that the meridian-oriented labyrinth pattern can be used as a dial of a sundial-calendar. To do this, a gnomon, an object that casts a shadow, must be installed in the center of the labyrinth. The area covered by the shadow in one year has the shape of a labrys. Geographical methods of heritage research open up new possibilities: they allow us to consider the connection of ancient signs and images with the measurement of space and time; they can be a source of information about the tilt of the earth's axis and gravity. It is advisable to create a geodatabase on the labyrinths of Europe.

Keywords Republic of Karelia · Stone labyrinths · Prehistoric cultural heritage · Sundial-calendar · Geographical methods · GIS

Introduction

Ancient labyrinths are spiral or concentric arrangements of wild stone (up to 40 cm in diameter) or petroglyphic drawings. Labyrinths are found on all continents except Antarctica, and the largest number is concentrated in Northern Europe. Labyrinths are located near waterways, permanent and temporary settlements—on islands, river banks and seas; they are found both on plains and in the mountains. Most labyrinths

A. Paranina (✉) · R. Paranin
Herzen State Pedagogical University of Russia, St. Petersburg, Russia
e-mail: galina_paranina@mail.ru

R. de Miguel González and P. Rivero Gracia (eds.), *Geospatial Technologies for Heritage Education*, Key Challenges in Geography,
https://doi.org/10.1007/978-3-031-96372-8_10

date back to the Neolithic and Bronze Ages. The oldest image was found in Siberia—18,000 years old. According to shape, labyrinths are classified as round, oval and square; according to structure, they are mono- and bi-spiral, concentric and radial (Kern 2007; Kuratov 2008).

The ancient labyrinths of Europe became a mystery back in ancient times. Thanks to the myth of the Minotaur, they are associated with a confusing trajectory of movement and danger to life. Mythological plots are still perceived literally, without an attempt to understand the meaning of a subtle allegory. According to one of the ancient Mediterranean legends of the world, creation begins as follows: with one side of the labrys axe God cuts through the darkness and receives light, and with the other side of the axe he cuts himself and is surprised to find a torch with fire in his hand. The plot is perceived as tragic, but in fact it describes not self-sacrifice, but the simultaneous birth of light and shadow—while the double-sided two-horned axe personifies their genetic relationship. Geographical research methods allow us to restore the primary rational functions of labyrinths as instruments for orientation by the sun's shadow and understand the language of the creators of ancient legends.

The theoretical basis for reconstructing the practical purpose of labyrinths as dials of sundial calendars is general physical geography (section "Earth as a Planet"). Geospatial analysis includes: (1) field research—photographs in plan, observations and measurements, (2) clarification of the position of objects—comparison of plan images (from a quadcopter) with space images, (3) calculations and modeling of the position of the solar shadow in the labyrinth (astrocalculators and other software are used). At the same time, the use of publications in the humanities area and photographs from the Internet is problematic, since the basis for reconstructing the instrumental functions of objects is precise orientation and geolocation (Paranina 2010, 2012, 2016, 2017, 2020, 2025; Paranina and Paranin 2015, 2017a, b, 2018; Paranina et al. 2017).

The author's concept of the practical purpose of labyrinths as tools for orientation by the Sun was formed in 2008–2019 at the White Sea sites. The concept is based on the following known facts: (1) solar navigation has no alternative in the conditions of polar days and white nights, (2) ancient seafaring traditions are illustrated by petroglyphs of the White Sea created 6000 years ago (boats with a capacity of 10–20 people, scenes of whaling from several vessels, etc.). In the course of field research, accurate data on the spatial characteristics of the objects were obtained. Observations and calculations of the position of the solar shadow in the labyrinth gave positive results. In response to the publication of the results, confirmations were received from different regions of Europe: (1) in the North Caucasus in Russia, the functions of the dials were reconstructed for images of seven-arc labyrinths (Israpilov 1993; Khetagurov 2004); (2) on the petroglyphs of Mogor in Spain, compositions of several labyrinths perform the functions of a calendar (the mathematics teacher Jose Luis Galovart, using the shadows of vertical objects, established a connection in the arrangement of the images of the labyrinths and the illumination of the Earth on the days of the equinoxes); (3) the Mostishchesky labyrinth on the high (60 m) bank of the Don River has been interpreted by archaeologists as a solar calendar—azimuths measured from the central pile of chalk debris to the granite blocks correspond to the position of the Sun at noon, as well as on the days of the equinoxes and solstices; under the supervision of professors of the Voronezh State University A.T. Sinyuk and V.D. Berezutsky, this work was completed by a student (Chekmenev 2001). Our calculations have shown that the pile of stones in the center of this labyrinth is convenient for installing a gnomon, and the midday shadow of a 3.76 m high pillar moves from the inner arc of the labyrinth to the outer one during the year—the concentric circles of the labyrinth act as a time scale (Paranina 2012). The listed results were obtained by the researchers independently, so they can be considered as independent confirmation of the thesis: the labyrinth is a sundial-calendar. The main reason for the oblivion of the primary rational function of labyrinths (and the instrumental functions of many other heritage sites) is the progressive development of orientation technologies: by the beginning of the Bronze Age, models of the world had formed that included the main elements of the celestial sphere—an example is the yurt of nomads, which served as a small astronomical observatory (Mashanov 1985); clay tablets of ancient Babylon with calculations of the length of a person's shadow for each month of the year for 7 climates—latitudes of the northern hemisphere (Van der Waerden 1974); sailors and scientists of antiquity used a celestial sphere cast in metal—an armilla; the widespread practical use of the celestial sphere before our era is reflected in numerous spherical sundials of various types (Tadiħ 2002).

The reason for the sacralization of navigational cultural monuments is that knowledge of space-time is of decisive importance for the survival and development of society. It ensures: (1) the effective development of the territory and its resources; (2) rational behavior—the synergy of human activity with the regime of solar energy inflow to the earth's surface; (3) the consistency of the traditional model of the world (worldview) with stable geographical processes and the laws of the Universe.

Objects and Region of Research

The ancient labyrinths of Europe are located mainly on the coasts of the northern seas—from the Isles of Scilly in Great Britain to Archipelago Novaya Zemlya in Russia. The preservation of objects in the Arctic part of the area was facilitated by the harsh natural environment, which limits the rate of natural overgrowing with turf, the possibility of intensive economic activity and the danger of anthropogenic degradation of the landscape. The first mention in historical documents of Russia in the fourteenth century refers to labyrinths at the mouth of the Kola River and on the shore of the Varanger Fjord. In the nineteenth century, researchers discovered numerous labyrinths in the interior of the Kola Peninsula, but they were soon destroyed by the Sami and the incoming Slavic population (Baer 1844; Eliseev 1883).

To the south, stone spirals of labyrinths are found in cultural layers of the Bronze Age (the Mastishche-1 settlement, the middle reaches of the Don River). In many regions of Europe, images of labyrinths have been preserved: in the form of drawings on rocks (Mogor on the Iberian coast, Val Camonica in the Alps, the labyrinths of Dagestan and Alania in the North Caucasus); in wood carving and embroidery on fabric, in the ornaments of everyday and religious objects (Etruscan vases, Roman mosaics, manuscripts, decoration of medieval temples). In the traditional culture of all peoples of the world, labyrinths are considered a group of solar signs with positive life-affirming content. In the Christian tradition, labyrinths are reinterpreted as a difficult path of sin and repentance.

Archaeologists' opinions on the purpose of labyrinths vary greatly and usually assert mystical thinking and irrational behavior: a place of communication with giants or gnomes; a form of symbolic burial; a fishing trap or its scheme; simulators for adjusting the vestibular apparatus before going out to sea (excursion materials of the Solovetsky Museum-Reserve). At the beginning of the twentieth century, astronomer Svyatsky (1913) in Russia and ethnographer Hermann Wirth in Germany came closest to interpreting labyrinths as astronomical instruments. However, they were unable to reconstruct the observation algorithm, since they relied on the technologies of near-horizon observatories and considered the labyrinth to be a direct projection of the sky. And only the use of reverse sighting on the gnomon of the sundial-calendar showed that the labyrinth is a dial, and its meaning is filled with the arrow of the sun's shadow.

About 50 stone labyrinths have been preserved on the coast of the White Sea. Most of them are located near the sites of seasonal fishing. Individual objects have been preserved in the settlements of the Pomors (Ponoy). A unique cluster of labyrinths is located on Bolshoy Zayatsky Island of the Solovetsky Archipelago (30 objects on an area of 1.36 km^2), which is located on the ancient waterway from the village of Kem to the Onega Peninsula (Figs. 1 and 2). According to archaeological data, the labyrinths date back to the Neolithic and Bronze Age (5000–3000 years ago). The White Sea petroglyphs (over 6000 years old) help to imagine the activities of ancient man in the region: skiers, deer, large ships, scenes of sea and forest hunting.

In the conditions of polar days and nights, sea navigation on the White Sea is possible only in the summer half of the year—between the days of the spring and autumn equinoxes. The season of effective salmon fishing lasts from May to August.

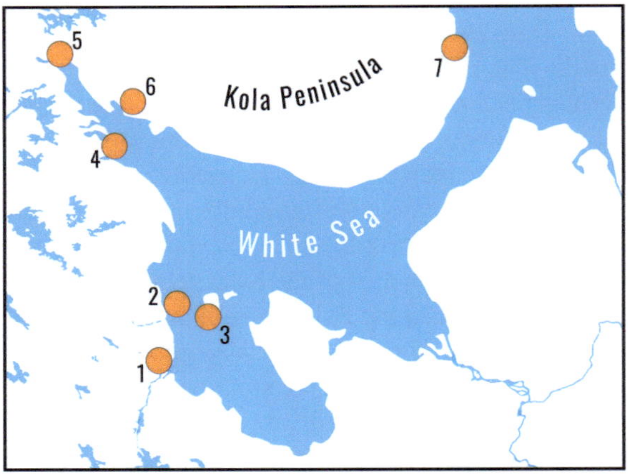

Fig. 1 Geography of research objects on the shores of the White Sea. (1) Belomor petroglyphs, (2) Labyrinths of the Solovetsky Arch, (3) Labyrinths of the Kuzov Arch, (4) Labyrinths of the Keretsky Gulf, (5) Labyrinth of the Kandalaksha Gulf, (6) Labyrinth near Umba, (7) Labyrinth in the village of Ponoy

Fig. 2 Labyrinths of the White Sea: **a**—Oleshin Island, Kuzova Archipelago; **b**—Labyrinth No. 1 of Bolshoy Zayatsky Island, Solovetsky Archipelago; **c**—Stone plate in the center of Labyrinth No. 3 (convenient for determining true noon). Numbering by N.N. Vinogradov (1927)

To this day, local Pomor residents and fishermen from neighboring regions go out for seasonal fishing at this time. In addition, the route to the shores of the Arctic Ocean has passed through the White Sea region since time immemorial. Thus, Elias Lönnrot (collector of the Karelian epic "Kalevala") wrote in his travels on February 6, 1837: "Starting from the middle and until the end of May, convoys stretch in a continuous stream through Keret. From the environs of Kem, from Suma, Shunga, etc. people go—the so-called Murmansk people—to Kandalaksha and from there to the Arctic Ocean to go fishing" (The Travels of Elias Lönnrot: …, 1985).

Geospatial Analysis Methods

Geospatial analysis of ancient cultural heritage sites must be carried out at the local, regional and global levels. At the local level, the range of capabilities of a specific tool is determined; at the regional level, the structural features associated with the illumination regime at different latitudes are highlighted; at the global level, universal elements of technologies that are invariant at any point in geographic space are identified.

We will consider *the analysis of the individual structure of labyrinths* using the example of the labyrinth of Bolshoy Zayatsky Island—No. 1, according to the numbering of the first researcher Vinogradov (1927).

Instrumental labyrinth functions can be set on the basis of the correspondence between the elements of the spatial structure and the parameters of the lighting mode (altitude and azimuth of the sun). Calculating the height and installation space of the gnomon, in which the shadow end positions coincide with inner and outer diameters of the arcs of the labyrinth, was made using the original equation based on the trigonometric identity ($H = \text{tg}\ \alpha \bullet \text{A}$). Both parts of the equations give one result—the height of the gnomon and describe the relationship of the angle of incidence of sunlight and the length of the shadow of α A: left for summer, right for the winter (1). By the distance measured from the center to the extreme edges of the labyrinth (in this example, 1 and 5.5 m), added an amendment, allowing to clarify its position in the central addition:

$$\text{tg}\ 48.47(1 + x) = \text{tg}\ 4.97(5.5 + x) \tag{1}$$

The functional spatial characteristics of the instrument can be obtained as a result of calculating the correction **x** to the distances measured to a conditionally selected center:

$$\begin{aligned}
&1.13(1 + x) = 0.09(5.5 + x)\\
&1.13 + 1.13x = 0.5 + 0.09x\\
&1.13x - 0.09x = 0.5 - 1.13\\
&1.04x = -0.63\\
&x = -0.63/1.04\\
&x = -0.61
\end{aligned}$$

Adjusting the distances to the center gives the size of the shadow in different seasons (m):

$$\begin{aligned}
&1 + x = 1 - 0.61 = 0.39\\
&5.5 + x = 5.5 - 0.61 = 4.89
\end{aligned}$$

The height of the gnomon is determined using the same Eq. (1) (m):

$$1.13 \cdot 0.39 = 0.09 \cdot 4.89$$
$$0.44 = 0.44$$

The greatest number of functions (compass, clock, calendar) perform bi-spiral labyrinths, symmetric with respect to the meridian. In other aspects, solstice, equinox, and other dates are determined by the azimuth of sunrise/sunset.

Determination of the functions of all elements of the classical labyrinth pattern by comparing the radius of its arcs and the changes in the length of the midday shadow of the gnomon over the year (Fig. 3) using the example of labyrinth No. 1 of Bolshoy Zayatsky Island: the first and second arcs act as markers of the summer solstice and equinoxes, the northern addition serves as a screen for the longest shadow (December–January). In latitudes where the winter shadow is shorter, this function is performed by the outer arc. The entrance to the labyrinth is the beginning of the annual cycle or fishing season. Rhythmically repeating elements of the labyrinth serve as a ruler that is filled with content in the specific conditions of geographic space, landscape development, and season of the year.

The movement of light and shadow of the gnomon reveals the instrumental functions of all elements of the labyrinth design. In the daily movement, the shortest shadow is observed at midday and points to the geographic north, this allows using a drawing oriented along the meridian as a compass and a clock (the central addition

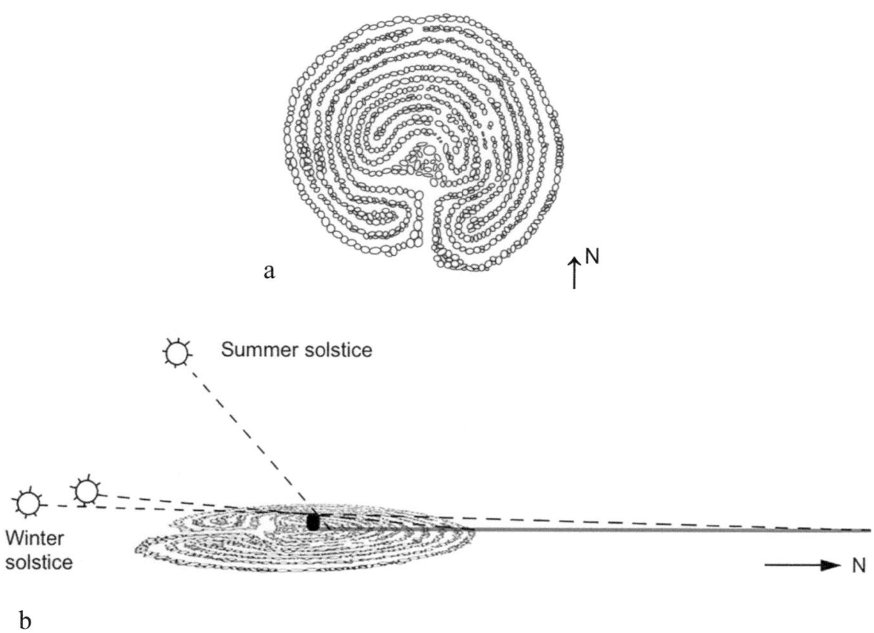

Fig. 3 Labyrinth No. 1 of Bolshoy Zayatsky Island: **a**—plan of the labyrinth (d = 10 m); **b**—changes in the length of the midday shadow of the vertical gnomon (the vertical and horizontal scales are the same)

serves as a gnomon). The shadow of a vertical object moves quickly in the morning and evening, and slowly in the middle of the day, therefore, long before our era, the tilt of the gnomon to the North Star was used to divide the day into equal parts, in which the gnomon is parallel to the earth's axis, and the speed of the shadow exactly corresponds to the speed of daily rotation. However, there is every reason to assume that in the labyrinth the gnomon was installed vertically, and the main function of the labyrinth drawing is calendar. This is indicated, in particular, by their location—as a rule, far from permanent settlements, in places of seasonal trades.

During the year, the midday shadow moves unevenly: when the Sun is high in the summer—slowly, when the Sun is low in the winter—quickly (Fig. 4, rows—1 and 2). The first arc from the center corresponds to the highest position of the midday Sun, then it gradually decreases, occupying the same position in May and July, April and August, March and September, February and October, January and November. In December, the shadow goes far beyond the observation site. In more southern latitudes, the Sun does not descend so low, and the outer arc of the labyrinth is a marker of the winter solstice.

The relationship between the length of the shadow and the length of the shadow can be analyzed in more detail using the graphical model presented in Fig. 5. The graph of the annual dynamics of the midday shadow (Fig. 5a) shows a significant difference between the three winter months (1, 11 and 12); February (2) and October (10) are transitional in nature, the marks of the summer months hardly change in length. Excluding the longest midday shadow (12—December) from the graph allows us to zoom in and examine the summer months (Picture 5b). It is evident that in the light

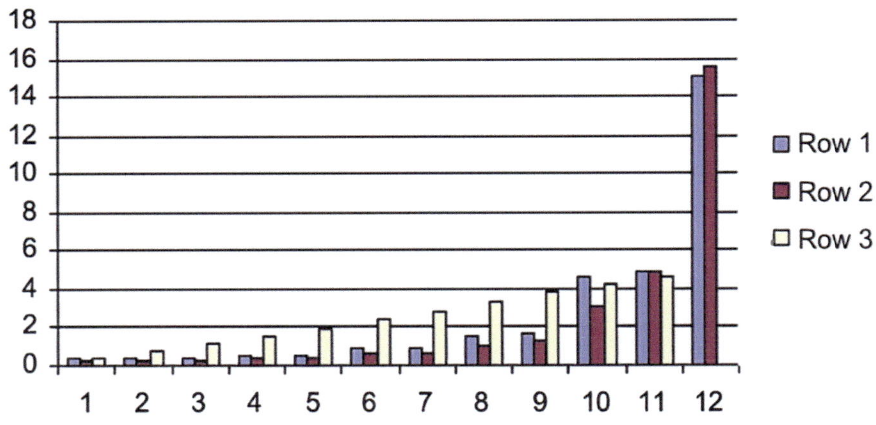

Fig. 4 Labyrinth No. 1, B. Zayatsky Island, length of midday shadow of vertical gnomon in relation to diameter of arcs: Row 1—in 2010; Row 2—in 5000 years ago, Row 3—arc diameters. Vertically—dimensions (m) measured from the center. Horizontally—numbers: arc numbering—from the center; month numbers in order of increasing midday shadow length (1—summer solstice, 2 and 3—May and July, 4 and 5—April and August, 6 and 7—March and September, 8 and 9—February and October, 10 and 11—January and November, 12—December)

half of the year the midday shadow hardly changes in length and is limited by the radius of the arc located near the center.

Comparison of graphs constructed for different times (here 2010 and 5000 years ago) shows that the calendar function of the labyrinth can be carried out for a long time without significant changes. This explains the long-term preservation of the tradition of building labyrinths, the stability of using the solar sign in various cultures. In regions where the Sun does not sink low, midday shadows are shorter, and the

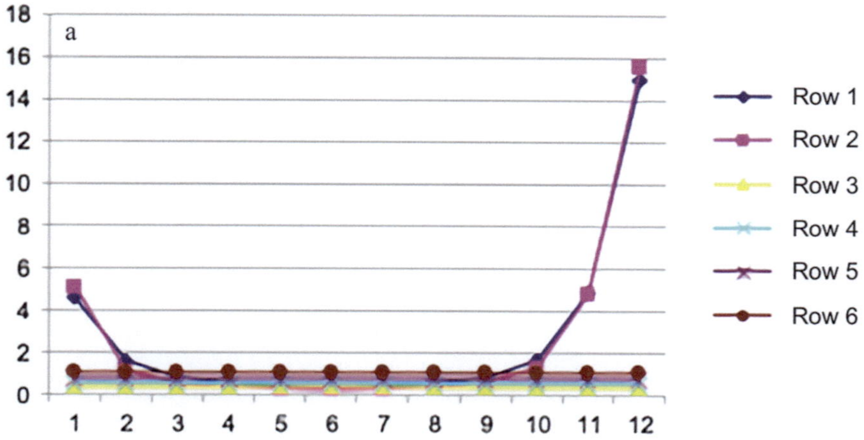

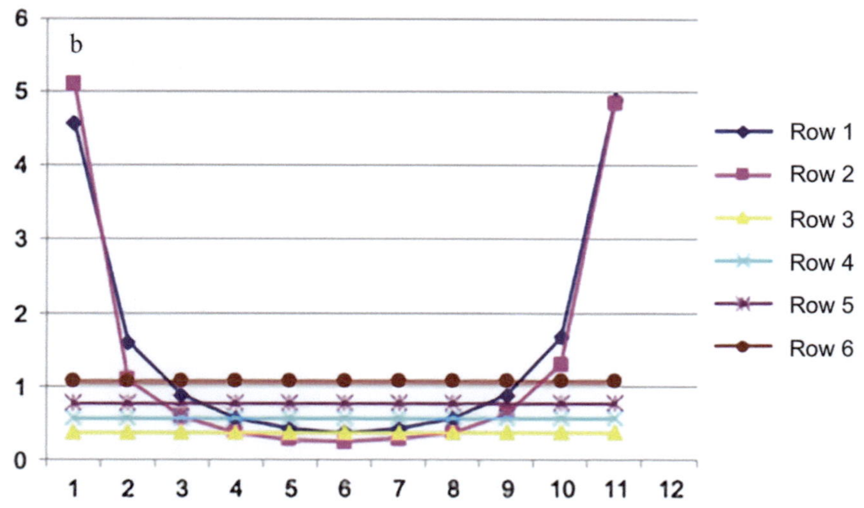

Fig. 5 Annual changes in the length of the midday shadow in relation to the two inner arcs of the labyrinth: **a**—for 12 months (in series); **b**—without December. Row—1 according to data for 2010, Row—2 according to data for 5000 years ago, Rows 3 and 4—the first arc of the labyrinth from the center, Rows 5 and 6—the second arc of the labyrinth from the center. Horizontally—the numbers of months in order, vertically—distances (m)

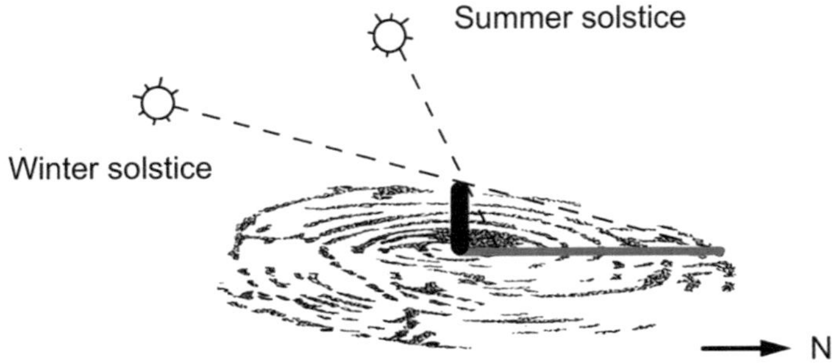

Fig. 6 Reconstruction of the Mostishchensky labyrinth 51.5° N. The height of the gnomon is 3.76 m. The vertical and horizontal scales are equal

outermost arc corresponds to the winter solstice. An example is the Bronze Age labyrinth located at a latitude of 51.5°, near the village of Mostishche, in the Voronezh region. Concentric chalk layouts were discovered by archaeologists under a layer of fine earth and turf (60 cm) on a ledge of the high bedrock bank of the Don River (60 m above water level) (Fig. 6).

In classical labyrinths, 7 arcs provide a count of 12 months, since 5 arcs of shadow pass twice—in the first and second halves of the year. This form reduces the costs of materials and labor, but preserves the idea of harmony of two serpentine spirals reflecting the ascending and descending movement of the Sun.

The unevenness of the annual dynamics of the midday shadow and the variability of weather conditions have important consequences: (1) maintaining direct sighting; (2) taking into account the position of the zodiacal constellations; (3) using the cycle of the Moon. To count the days and phases of the Moon in the classical bi-spiral labyrinth, it is convenient to use its 7 arcs, including in the count 8-day quarters to equalize the fractional duration of the phases of the Moon (lunar months last 28–29 days).

A comparative analysis of the labyrinths showed that rhythmically repeating elements serve as a ruler/scale, which is filled with real content in specific conditions of geographic space and time. This is based on a deep tradition of using rhythmic structures for counting: natural and artificial (notches, geometric ornaments of repeating graphic elements). The diversity of tools is reflected in the logical formula "technologies are universal, objects are unique".

Spatial analysis of the area covered by the gnomon shadow

As is known, the basis of the dial (limb/kadran) marking of the sundial calendars is formed by drawing shadows diverging from the point of installation of the gnomon. However, spatial analysis of the shapes of the area covered by the shadow gives new interesting results. The sum of the daily shadows of the gnomon has the shape of a fork, when drawing the shadow of individual rays it resembles the tail fin of fish, the wings of a bird or the horns of a bull—in them one can recognize the attributes of the ancient gods who control time.

The graphic model of the year (radar chart in Excel) based on the calculations of the sum of daily shadows (Fig. 7a) with an interval of 1 h on the 22nd of each month resembles a labrys (Fig. 7b)—a double-sided two-horned axe of the gods of darkness and light. At the same time, the reconstruction of the meanings of the labrys moves away from the established definition of "sacrificial axe" in cultural studies. The following are filled with solar meaning: A. Evans's definition of "labyrinth - the house of the labrys" (about the Knossos labyrinth palace). The semantics of the image of the "golden hatchet" preserved in folk tradition (in the text of carols performed on the days of the winter solstice) becomes clear. The foundations of time images are revealed in a new way: the lotus is a symbol of the "day of the gods" (the day of the gods lasts a year); the many-armed or six-armed Shiva (the six directions correspond to the equinoxes and solstices), etc.

The technologies of the labyrinth-gnomon can serve as the basis for a new reading of ancient mythological stories: (1) *the spear and rope* depicted next to the labyrinth on ancient coins *are necessary and sufficient materials for creating a tool* for orientation in space–time (the spear is a gnomon, the rope is a universal basis for marking the pattern: a compass, a measuring tape, a knotted "Egyptian triangle", a flexible model of the outlined lines, a plumb line); (2) the labyrinth is similar to the "apple of knowledge" in that it allows you to measure time, *depriving you of the illusion of eternity,* snakes and the world tree—like the labyrinth-gnomon—a geocentric model of the Universe; (3) One-eyed Odin is a gnomon that "sees" all points in the sky, is *a projection of the earth's axis and the axis of the world in the celestial sphere*. His sons

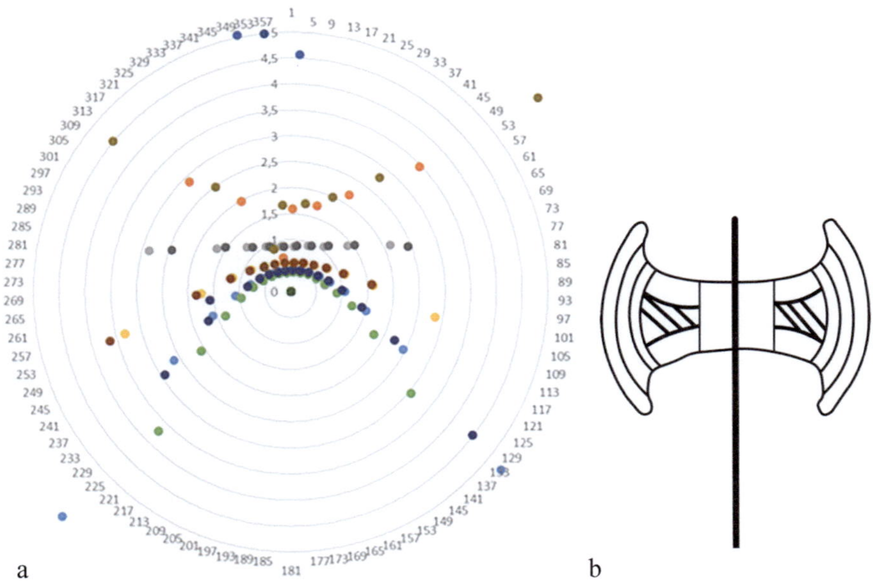

Fig. 7 Graphic and figurative models of sunlight: **a**—area covered by the gnomon's shadow over the year (graph for the 22nd of each month); **b**—golden labrys, an exhibit in the Heraklion Museum

Thor and Loki are personifications of the reliable reference point to the north and the elusive shadow; Thor's hammer is the sum of shadows over the year (analogous to the labrys), because: it cannot be lifted (shadow), it always returns (the midday shadow approaches the object and points north) and has all-destroying power (time).

Regional differences in the structure of the labyrinth pattern are associated with the peculiarities of the lighting regime at different latitudes (Table 1).

Table 1 shows the change in the azimuths of sunrises on the days of the solstices at different latitudes (the azimuths of the sunsets are located in the western half of the horizon—symmetrically to the sunrises relative to the midday line). In the Arctic, this indicator is absent—the days of the solstices fall on polar days and nights; in subpolar latitudes, the azimuth of the sunrise changes at a high speed; in middle latitudes, the azimuth changes are gradual, in tropical latitudes—minimal. It is noteworthy that at the latitude of the Arctic Circle, the points of sunrise and sunset divide the horizon line into 360 segments during the year. And at a latitude of 55.5°, the directions of sunrise and sunset on the days of the solstices form a regular oblique cross—this allows you to create an ideal drawing of a double-spiral labyrinth, in which the main and intermediate sides of the horizon are clearly marked.

Comparison of the length of the midday shadow at different latitudes shows that the optimal conditions for calendars based on the change in the length of the shadow are the temperate zone of illumination (between the tropic and the polar circle). Here, the shadow moves away from the center to the periphery for half a year and then returns; the outer boundary of the labyrinth corresponds to the length of the shadow at the winter solstice (or in the months before and after it); the inner arc from the center corresponds to the summer solstice. In the cold zone, the shadow is absent during the polar night; in the hot zone, the midday shadow disappears twice a year—when the sun reaches the zenith point. In addition, in summer, the midday shadow falls towards the south pole for some time: from one day (slightly below the tropic) to 6 months (at the equator).

The method of constructing circular petal diagrams (in Excel) allows you to represent the marking of the site with circles, with an interval of one month (Figs. 8 and 9). The diameters of the circles, calculated based on the altitude of the Sun on the 22nd of each month, are extremely unevenly distributed at the latitude of the polar circle and near it (Fig. 8). The most uniform distribution of radii is characteristic of latitudes of 30–50° (Fig. 9).

Spatial analysis helps to analyze the actual data on labyrinths obtained from different sources. Calculations for 7 arcs of the Makhchesky labyrinth (42°, Republic of North Ossetia–Alania) show significant differences in the duration of months. This is consistent with ethnographic evidence for the peoples of the Caucasus, for whom the duration of time periods in the annual economic (phenological) calendar varied from 2 weeks to 2 months (Khetagurov 2004).

The ratio of the considered indicators is analyzed and reflected in the combined graph for the latitudes of the temperate zone (Fig. 10).

The azimuth of sunrise at the summer solstice varies from 64° on the tropics to 0° on the Arctic Circle. In the polar region, it is not determined, since the Sun does not go below the horizon at this time (Fig. 10, Row 2).

Table 1 Annual changes in the azimuth of sunrise, the height of the Sun above the horizon and the length of the midday shadow for an object $H = 1$ m in the northern hemisphere (0–70° N)

Latitude (°) (´)	Azimuths of sunrises (°)			The height of the Sun above the horizon (°) *Length of midday shadow, m*											
	WS	S/AE	**SS**	1	2	3 SE	4	5	6 SS	7	8	9 AE	10	11	12 WS
70	no	90	**no**	0.25 *250*	9.69 *5.88*	20.50 *2.70*	32.09 *1.59*	40.32 *1.18*	**43** *1.05*	40.34 *1.18*	31.88 *1.61*	20.44 *2.70*	9.07 *6.25*	– 0.06 –	– 3.44 –
66°33´	180	90	**0**	3.7 *16.67*	13.14 *4.35*	23.95 *2.27*	35.54 *1.41*	43.77 *1.04*	**47** *0.94*	43.79 *1.04*	35.33 *1.41*	23.89 *2.27*	12.52 *4.55*	3.39 *16.67*	0.01 *10.000*
65	160	90	**20.03**	5.25 *11.11*	14.69 *3.85*	25.50 *2.08*	37.09 *1.32*	45.32 *0.99*	**48** *0.89*	45.34 *0.99*	36.88 *1.33*	25.44 *2.08*	14.07 *4.00*	4.94 *11.11*	1.56 *33.33*
60	142.86	90	**37.40**	10.25 *5.56*	19.69 *2.78*	30.50 *1.70*	42.09 *1.11*	50.32 *0.83*	**53** *0.74*	50.34 *0.83*	41.88 *1.11*	30.44 *1.70*	19.07 *2.86*	9.94 *5.56*	6.56 *8.33*
50	128.41	90	**51.82**	20.25 *2.70*	29.69 *1.75*	40.50 *1.18*	52.09 *0.78*	60.32 *0.57*	**63** *0.5*	60.34 *0.57*	51.88 *0.79*	40.44 *1.18*	29.07 *1.79*	19.94 *2.78*	16.56 *3.33*
40	121.29	90	**58.74**	30.25 *1.72*	39.69 *1.21*	50.50 *0.83*	62.09 *0.53*	70.32 *0.36*	**73** *0.11*	70.34 *0.36*	61.88 *0.54*	50.44 *0.83*	39.07 *1.23*	29.94 *1.72*	26.56 *2.00*
30	117.39	90	**62.74**	40.25 *1.18*	49.69 *0.85*	60.50 *0.57*	72.09 *0.32*	80.32 *0.17*	**83** *0.11*	80.34 *0.17*	71.88 *0.33*	60.44 *0.57*	49.07 *0.87*	39.94 *1.19*	36.56 *1.35*
23°27´	115.80	90	**64.70**	46.80 *0.95*	56.24 *0.67*	67.05 *0.42*	78.64 *0.20*	86.87 *0.06*	**90** *0.00*	86.89 *0.05*	78.43 *0.21*	66.99 *0.42*	55.61 *0.69*	46.49 *0.95*	43.11 *1.06*
20	115.05	90	**64.97**	50.25 *0.83*	59.69 *0.59*	70.50 *0.36*	82.09 *0.14*	**90** *0.006*	86.56 *0.06*	**90** *0.006*	81.88 *0.14*	70.44 *0.36*	59.07 *0.599*	49.94 *0.83*	46.56 *0.94*
10	113.85	90	**66.19**	60.25 *0.57*	69.69 *0.37*	80.50 *0.17*	**90** *0.04*	79.68 *0.18*	76.56 *0.24*	79.66 *0.18*	**90** *0.03*	80.44 *0.17*	69.07 *0.38*	59.94 *0.58*	56.56 *0.66*
0	113.44	90	**66.56**	70.25 *0.36*	79.69 *0.07*	**90** *0.01*	77.91 *0.21*	69.68 *0.37*	66.56 *0.43*	69.66 *0.37*	78.11 *0.21*	**90** *0.01*	79.07 *0.19*	69.94 *0.37*	66.56 *0.43*

* Abbreviations: WS—winter solstice, SS—summer solstice, SE—spring equinox, AE—autumnal equinox

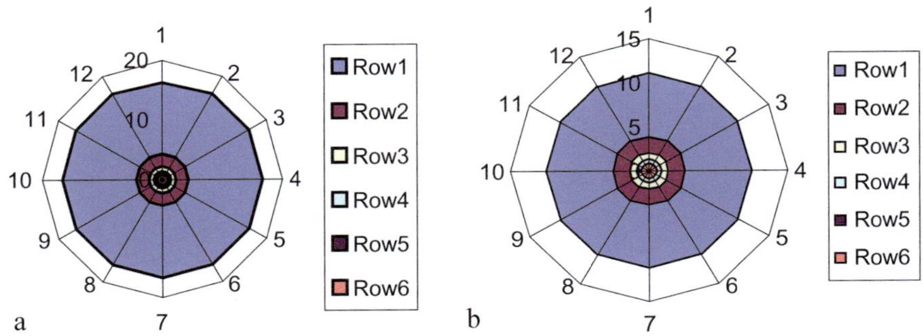

Fig. 8 High-latitude labyrinth models: **a**—for the Arctic Circle 66.5° N; **b**—for Zayatsky Island 65° N; Row 1—1 and 11 months; Row 2—2 and 10; Row 3—3 and 9; Row 4—4 and 8; Row 5—5 and 7; Row 6—June. Winter—Rows 1 and 2; Equinox—Row 3; Summer—Rows 4–6. December (month 12) excluded for larger scale images

The relationship between the height of the gnomon and changes in the length of the midday shadow at different latitudes by season: (1) during the polar day, the shadow has the form of a closed elliptical structure (a ring-shaped one is possible only at the pole)—the "ankh" type; (2) in subpolar latitudes, it is difficult to fix the midday shadow in winter—the "Babylon" type; (3) at a latitude of 50–60°, the midday shadows over the year fit into a circle of small diameter—the "classical labyrinth" type; (4) at 40–30°—the radius of the outer circle is twice the height of the gnomon—the "staff" type (≈ 2 m); (5) in the tropics, all the shadows fit on the gnomon, as on a ruler, but the edges for different seasons are distinguished—the "rod"/"scepter" type (Fig. 10, Row 1).

The ratio of the lengths of the midday shadow at the summer and winter solstices varies from 1:6.4 at latitude 45° to 1:100 and more at the boundaries of the temperate zone. At latitudes 51° and 39°, it is 1:7, which corresponds to the height of a standard staff (6–7 feet) (Fig. 10, Row 3).

Based on the analysis, it is possible to identify key parallels and zone the territories of the temperate zone of Europe and the whole world. It is also worth noting the historical facts: parallels 51.4° and 38.6° coincide with the boundaries of the *seven climates* identified in the texts of the cuneiform tablets of Ancient Babylon—n (90°/7); parallel 45° is mentioned in sources of ancient times as Great Media along with the Great Boundary of the Parts of the World (0° meridian—Alexandrian).

Zoning the land according to the conditions of solar navigation allows us to interpret the shape of not only labyrinths, but also other ancient heritage sites, including petroglyphs (Paranina and Paranin 2017a). The map also shows the areas of the Earth in which it is possible to predict the finds of heritage sites with similar properties.

The global invariant of solar navigation is determined by the rotation of our planet in the flow of sunlight: sunrises, sunsets and upper culminations of the Sun in a daily cycle, the change of astronomical seasons—in an annual cycle. Constants in the illumination mode of all points on the earth's surface formed a system of stable and reliable signs: the shortest shadow at midday points to the pole (meridian), on the days

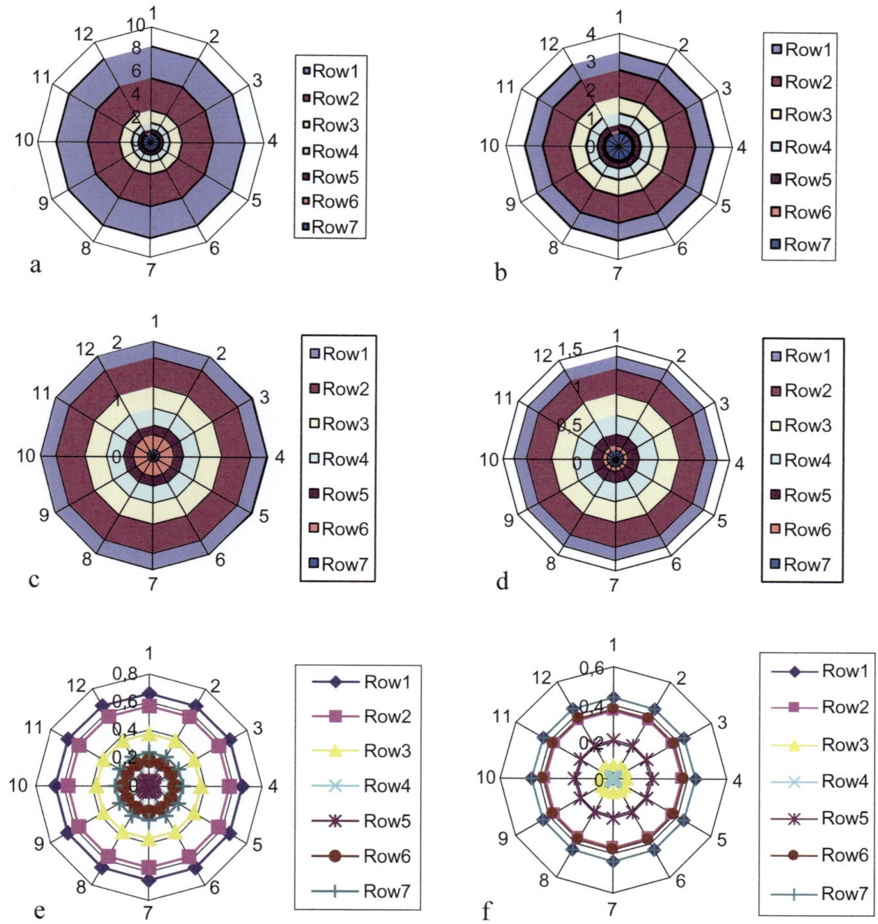

Fig. 9 Models of concentric calendars for middle and low latitudes: **a**—60° N; **b**—50° N; **c**—40° N; **d**—30° N; **e**—10° N; **f**—0°. Row 1—December; Row 2—January and November; Row 3—February and October; Row 4—March and September; Row 5—April and August; Row 6—May and July; Row 7—June

of the equinoxes the sun rises in the east (90°) and sets in the west (270°). The cross of the cardinal directions of the horizon became a symbol of life and success, the basis of ornaments, the layout of residential complexes and settlements, local and regional navigation networks, visual and toponymic marking of the territory. Determining the height of the Sun by the shadow of the gnomon allows the sphericity and size of the Earth to determine the latitude of the place and link individual points into a system, building a geocentric model of the world in the form of a celestial sphere. The widespread use of the gnomon labyrinth testifies to this level of development of prehistoric astronomy and geography.

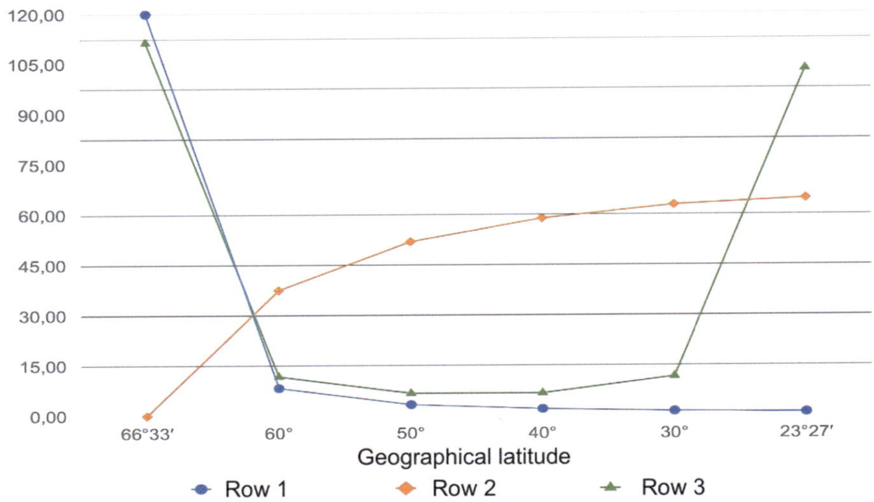

Fig. 10 Conditions of solar navigation in temperate latitudes: Row 1—ratio of the gnomon height and the length of the midday shadow; Row 2—azimuth of sunrise on the summer solstice; Row 3—ratio of the lengths of the midday shadow on the days of the winter and summer solstices (along the vertical axis—ratios of the lengths and horizontal angle of sunrise; along the horizontal axis—latitudes)

Ancient Labyrinths of Europe as an Archive of Information About Nature and Man

Dating of objects based on paleo-astronomical calculations. The change in the angle of inclination of the earth's axis and equator to the plane of the ecliptic with a period of 40,700 years (according to M. Milankovitch) is accompanied by a shift in the latitude of the tropics and polar circles: the tropics—from 21°58′ to 24°36′; the polar circles—from 68°02′ to 65°24′ (Fig. 11). These changes are reflected in the structure of the landscape and solar calendars, which is especially noticeable at the boundaries of the illumination belts (hot, moderate and cold).

Using the example of B. Zayatsky's labyrinth No. 1, we can see (Fig. 12) that in polar latitudes, along the course of long-period rhythms: (1) the height of the midday Sun, which determines the length of the midday shadow, changes insignificantly (Row 2)—this is convenient for long-term use of the calendar; (2) the azimuth of sunrise on the days of the solstices is very variable (Row 1)—this can be used to clarify the dating of objects.

The azimuth of the entrance to labyrinth No. 1 is 13°, which coincides with 6500 years ago (Fig. 13). This time of the labyrinth's creation is consistent with the age of the Belomor petroglyphs (more than 6000 years ago) and the Holocene climatic optimum (7000 years ago). However, the entrance to the labyrinth is most often located in the west (the labyrinth on Krasnaya Luda) or east (the labyrinth in Kandalaksha Bay and the large labyrinth on Oleshin Island). These azimuths

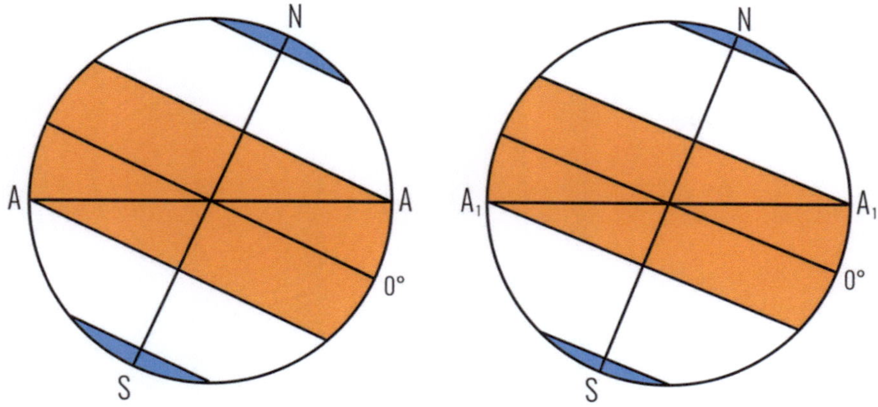

Fig. 11 The position of the polar circles and tropics depending on the angle of inclination of the earth's axis (NS) to the plane of the ecliptic A and A_1 [according to Kalesnik (1970)]

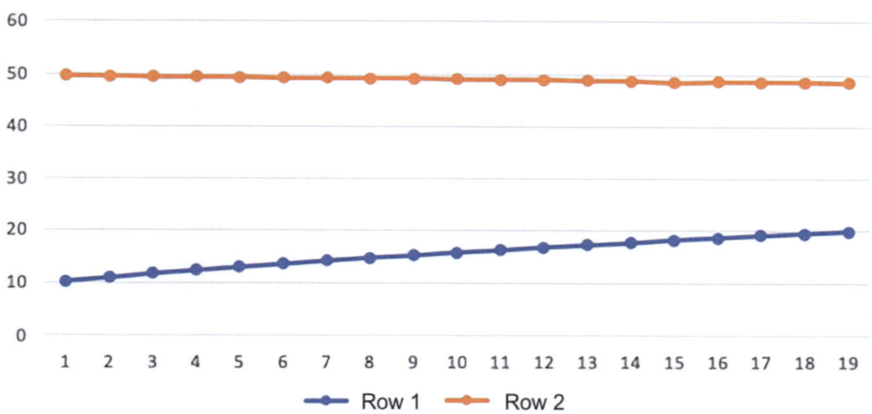

Fig. 12 Midday altitude (Row 1) and azimuths of sunrises at summer solstice (Row 2) at 65° N for 9000 years: vertical—angles (°); horizontal—time scale (500-year intervals, the passage of time—from left to right)

correspond to the days of the equinox at all latitudes, so they are not suitable for astronomical dating. The high frequency of such orientation may be due to the fact that labyrinths have been preserved in places of seasonal fishing, which was carried out exclusively in the light half of the year. Another orientation option was studied by us in the Umbian labyrinth: the entrance from the northwest corresponds to the azimuths of sunset at the beginning and end of the salmon fishing season (the duration of this season on the northern—"winter" coast of the White Sea: mid-May—mid-July). Obviously, the functions of labyrinths at the sites of seasonal camps are limited, since the fisherman spend the entire day at sea. A labyrinth that can be used as a

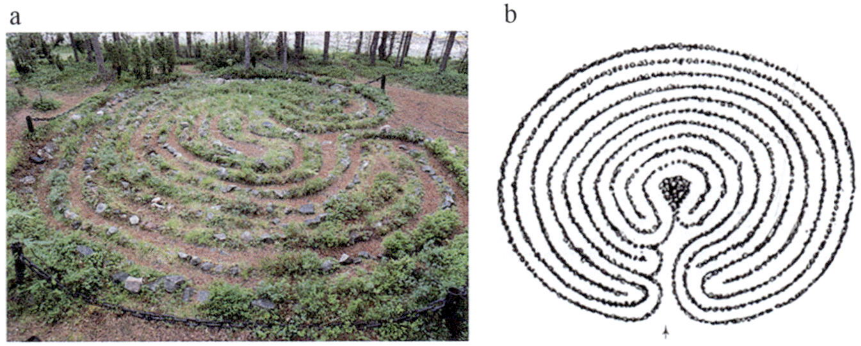

Fig. 13 Kandalaksha labyrinth: **a**—photograph from 2016, **b**—diagram by Durylin (1914)

sundial-calendar throughout the year has been preserved in the village of Ponoy (it is oriented along the midday line and is located near the house).

In general, the sources of uniqueness of labyrinths and other navigation objects are: latitude, position in the relief (absolute height and shape of the physical horizon, which deviates from the astronomical horizon in different ways everywhere), time of creation (planetary parameters—the tilt of the earth's axis and the displacement of the boundaries of illumination belts, the speed of axial rotation and the magnitude of gravity), phenological cycles, calendar traditions (the boundaries of economic seasons, the beginning of the countdown), the mode of use (permanent, seasonal), building materials, processing technologies, artistic design motifs.

It is also necessary to take into account the fact that some labyrinths were rebuilt during restoration. Thus, the modern drawing of the labyrinth in Kandalaksha does not resemble the scheme published in the early twentieth century (Fig. 13). In this case, a geospatial analysis of the object can only be carried out based on historical data and descriptions.

Possibilities of systemic research of ancient labyrinths. The uniqueness of astronomical instruments limits the area of their copying (borrowing), explains their widespread distribution, as well as the need for their network interaction. Modern geospatial research methods allow us to examine in detail each cultural heritage object and their entire set. The peculiarity of systemic research is in establishing the links between objects—among themselves and with the geographic environment (at different scales). This ensures the receipt of new information about the development of nature and culture in regions and on a global scale.

Labyrinths research algorithm includes the following steps: (1) The use of standard methods of description (measurement, comparison, metrological analysis—correlation with ancient measurement systems); (2) characteristics of the landscape (including the evolution of the climatic conditions during the Holocene geological and geomorphological features, including the dominant system in the fracturing of rocks and stretch lineaments—linear tectonic structures that are reflected in the landscape); (3) astronomical and calendar calculations of paleo-astronomical azimuths of sunrise/sunsets and the moon (to account for differences of physical and astronomical horizon), the height of the gnomon of a sundial-calendar and regulations midday shade seasonal (for geographical coordinates of the property); (4) The establishment of the correlations of the spatial characteristics of the object, of the landscape and important astronomical figures recorded at this point; (5) comparison of instrumental features of object with local and regional life support tasks in different historical eras (including the location of the object in the analysis of the transport communications system). (6) Analysis of the location of the object in the system of transport communications and social functions in different eras; (7) development of recommendations for rational use in science and tourism; determination of optimal conditions for preservation.

Comprehensive analysis of the purpose and use of labyrinths also includes the cultural context (neighboring archaeological sites, toponymy, linguistics, mythology, folk traditions). For systematic studies of the structure and evolution of the navigation network labyrinths of Northern Europe, it is necessary to create a database with the indication: (1) Geographical coordinates of the center points of labyrinths; (2) The true (geographic) entrance azimuths as well as directions to the point and linear drawing elements; (3) The values of the deflecting effect of the physical horizon (obtained using circular panorama); (4) Age of the relief elements (coastal terraces), on which there are labyrinths; (5) Dating labyrinths, derived from paleo-astronomical calculations.

Labyrinths as Objects of Cultural Heritage: Forms of Research Organization

Creation of a geographic database and GIS modeling. In the geospatial database, the first place is occupied by the geolocation and orientation of objects. Coordinates can be obtained using GPS or online mapping services (for example, Google Earth or Yandex Maps). To obtain accurate information about the orientation of objects, it is convenient to compare space images and planned photographs obtained from a quadcopter. As an example, let us consider labyrinth No. 3 of Bolshoy Zayatsky Island (Fig. 14). The photo was processed in the Pixelmator Pro program (transparency was reduced), which allowed it to be accurately combined with the details of the space image. It is evident that the labyrinth axis is oriented along the meridian, therefore, this is a typical sundial-calendar (due to the southern exposure of the slope, the plane of the labyrinth-dial is illuminated all year round). Some asymmetry in the arrangement of the elements of the drawing can be explained by the shape of the horizon, which controls the time of direct irradiation of the dial. However, this preliminary conclusion may be adjusted when compared with synchronous objects in the region in favor of regional or global factors.

The database for creating a GIS for the labyrinths of Europe should include the following key indicators:

Fig. 14 Clarification of the orientation of the labyrinth using the method of combining images obtained from a quadcopter and remote sensing

- altitude above sea level (gives an idea of the time when the coast was drained and the possibility of building labyrinths appeared; in the Baltic Sea in the Holocene, the maximum flooding dates back to the Litorina Sea stage, and the rate of drainage is associated with the tectonic uplift of Fenno-Scandia);
- geographical latitude: above or below the Arctic Circle (the classic bi-spiral labyrinth corresponds to the lighting regime in the temperate zone; beyond the Arctic Circle, it is an introduction, or its structure is transformed by the seasons of polar days and nights);
- orientation: along the meridian or other (an indicator of the purpose of creation—for permanent or seasonal use).

For ancient labyrinths oriented close to the meridional direction, the problem of determining the correlation between such indicators as the azimuth of the entrance to the labyrinth, geographic latitude, altitude above sea level and age can be solved.

Visualization of spatial information helps to analyze the processes of regional development. For the White Sea objects considered in the chapter (labyrinths and petroglyphs), a spatial model can be proposed based on three indicators: height, orientation, age (Fig. 15). The figure shows: (1) the movement of ancient man from the southern part of the sea to the northern (the climatic optimum of the Holocene did not occur simultaneously throughout the entire territory); (2) the long-term preservation of the tradition of building labyrinths on the western—Karelian coast (where there are still centers of fishing and yacht tourism); (3) the predominance in the northern part of the sea of objects of temporary use associated with places of seasonal fishing.

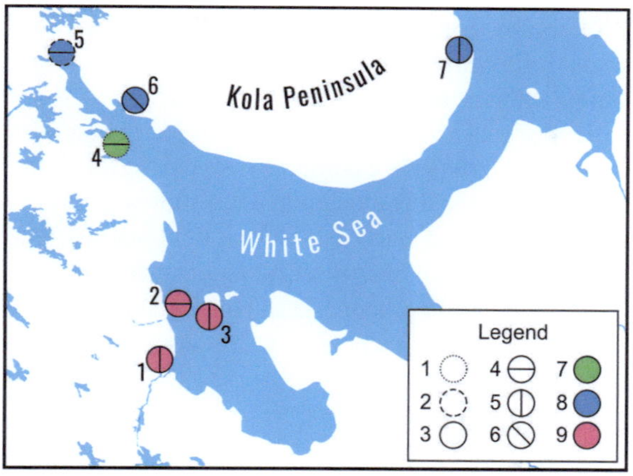

Fig. 15 Geospatial model of prehistoric cultural heritage sites of the White Sea with properties of navigation tools. Legend: 1 to 7—object numbers (see Fig. 1); 1–3 height above sea level (1—2 to 4 m, 2—5 to 10 m, 3—above 10 m); 4—6 orientation of the object axis (4—meridional, 5—latitudinal, 6—diagonal); 7—9 age according to archaeological data (7—less than 1 thousand years, 8—about 3 to 4 thousand years, 9—about 5 to 6 thousand years)

Geospatial analysis of cultural heritage sites allows you to organize work at three levels of cognitive activity: the reproductive level—action according to a model, the heuristic level—includes an independent choice of an algorithm, the creative level—involves independent formulation and solution of educational and research problems. Undoubtedly, the level of complexity of the tasks is determined by the degree of readiness of the researchers. Description and photo recording of objects (including shooting from a quadcopter), clarification of their orientation will be of interest to teachers and students of secondary school in the process of local history work or in the program of educational travels. Calculations of the position and height of an object based on a trigonometric function can be carried out in geometry lessons at school or in the first year of university when teaching general physical geography (section "Earth as a Planet"). The creation of a GIS for individual regions marks the transition to a creative level of cognitive activity of students and can be implemented by students when preparing their final qualifying works. The creation of a single database on the labyrinths of Europe can be a creative task of a temporary research team, including representatives of different countries (an international grant can facilitate the creation of such a team).

Conclusion

The goals of interdisciplinary research of cultural heritage sites are met by the navigation concept of information modeling of the world, which is based on the priority of human adaptation to the dynamic conditions of the geographical environment.

Geographical methods, using modern means of obtaining, storing, processing and analyzing spatial information, allow us to study the rational functions of ancient labyrinths associated with orientation in space and time.

For a systematic study of the labyrinths of Europe, it is necessary to create a geographical database, including their comprehensive description, as well as an indication of the exact orientation and geolocation. The following may participate in such research: university students and teachers—during field practices; secondary school teachers and students during local history work; independent researchers—travelers. The organizing role in the development and implementation of such a program can be played by the authors of the concept and the initiative link of Eurogeo.

References

Baer KE (1844) v. Ueber labyrinthförmige Steinsetzungen im Russischen Norden // Bull. l'Academie Imperiale de sciences de St.-Petersbourg 1:70–79

Chekmenev Y (2001) Astroarchaeological substantiation of the practical purpose of the Mostishchensky labyrinth // Mostishchensky complex of ancient monuments (Bronze Age—Early Iron Age). VSPU, Voronezh, pp 174–178

Durylin SN (1914) Kandalaksha "Babylon" (to study the northern labyrinths) report on a trip to the North. Pechatnya A. Snegireva, Moscow

Eliseev AV (1883) On the so-called Babylons in the north of Russia. News Geogr Soc XIX:12–16

Galileo G (1987) Assay master. Danilova YA, Nauka M (trans.), 272 p

https://yandex.ru/maps/2/saint-petersburg/?ll=30.314997%2C59.938784&z=11

https://planetcalc.ru/320/

Israpilov MI (1993) From Keger to Stonehenge: Dagestan's rock paintings tell their story. Science-Service, Makhachkala, 298 p

Kalesnik SV (1970) General geographical patterns of the Earth: textbook for the geogr. Mysl, Moscow, 282 p

Kern G (2007) Labyrinths of the world/translated from English by A. Rudakova, L. Shvedova; translated from Latin by N. Gorelova; translated from Italian and German by I. Khmelevskikh. Azbuka-Classic, St. Petersburg, 430 p

Khetagurov TN (2004) Archaeoastronomical properties of the Makhchesky labyrinth/rhythms of history. Publishing house of SOGU, Vladikavkaz, pp 113–126

Kuratov AA (2008) Stone labyrinths in the sacred space of Northern Europe. Pomor State University named after M.V. Lomonosov, Arkhangelsk, p 53

Laberintos de Mogor (Marín) jueves, 6 de diciembre de 2012. http://oestrymnio.blogspot.com.es/2012/12/laberintos-de-mogor-marin.html

Mashanov AZ (1985) Sundials in Kazakhstan. Sundials and calendar systems of the peoples of the USSR. Series "Problems of Universe Research" 10:103–113

Mysterious stone labyrinths in the Russian North—what are they? (PHOTOS). https://www.rbth.com/travel/334940-stone-labyrinths-russian-north?ysclid=m2ypesg5zz680463325

Paranina GN (2010) Light in the labyrinth: time, space, information. Asterion, St. Petersburg, p 123

Paranina AN (2012) Stone labyrinths in the system of ancient navigation. In: Divnogorsk collection: works of the Divnogorye Museum-Reserve, issue 3. Voronezh, pp 209–221

Paranina A (2016) Gnomon as source of information on planet rhythms. Int J GEOMATE 10(2):1815–1821

Paranina A (2017) Archaeological objects as elements of informational life support system and sources of information about the evolution of the environment. Int J GEOMATE 13(35):100–107

Paranina AN (2020) Space rhythms and technologies of astronomic navigation as factors of cultural genesis and sapientation. Int J Geomate 19(73):216–225

Paranina AN (2025) Navigation concept of information modeling of the world. Infra M, Moscow, p 271

Paranina AN, Paranin RV (2015) Northern labyrinths as navigation network elements. In: Activities in navigation: marine navigation and safety of sea transportation. London, pp 177–180

Paranina AN, Paranin RV (2017a) Primary navigation purpose of petroglyphs: reconstruction on the basis of the gnomon. Open Access Libr J 4(8):1–13

Paranina A, Paranin R (2017b) Information in geographical space as the basis of cross-disciplinary researches in culture geography. Eur J Geogr 8(3):67–77

Paranina A, Paranin R (2018) Exploration of geographical space—time as a factor of evolution of the nature and culture. Eur J Geogr 9(4):51–61

Paranina A, Paranin R, Khetagurov T (2017) Cross-disciplinary researches of objects of ancient heritage on the example of stone labyrinths and petroglyph. TransNav Int J Mar Navig Saf Sea Transp 11(4):729–734

Solovetsky State Historical, Architectural and Natural Museum-Reserve. https://solovky.ru/ru

Svyatsky DO (1913) Under the vault of the crystal sky: essays on astral mythology in the field of religious and folk worldview. Typ. M. Stasyulevich, St. Petersburg, IV, 188 p

Tadiħ M (2002) Sunčani chapels. Raduniħ, Beograd, 199 p

Travels of Elias Lönnrot: travel notes, diaries, letters, 1828–1842 (1985). Petrozavodsk

Van der Waerden BL (1974) Science awakening II: the birth of astronomy, 1st edn. Oxford University Press, 364 p

Vinogradov NN (1927) New labyrinths of the Solovetsky archipelago. Labyrinths of B. Zayatsky Island // Materials of the Solovetsky Society of Local History. Issue XII. Solovki

Collaborative Cataloguing of Cultural Heritage with GIS: Citizen Science Initiatives from the University of Zaragoza

Pilar Rivero, Iñaki Navarro-Neri, Silvia García-Ceballos, and Borja Aso

Abstract This chapter presents a comprehensive analysis of citizen science initiatives developed by the University of Zaragoza for the collaborative cataloguing of cultural heritage through Geographic Information Systems (GIS). These digital tools—Mural Hunter, CatPat, and RomanSites—enable users to geolocate, document, and reflect on tangible and intangible heritage elements. Designed within a framework of educommunication and the transition from ICTs to RICTs, these applications foster the formation of heritage cybercommunities and promote active, participatory learning. The chapter also explores the educational potential of these tools in formal and non-formal settings, offering didactic proposals aimed at engaging students in heritage preservation, critical analysis, and digital competence development. Through co-creation and the mediation of digital devices, this work demonstrates how GIS-based platforms can democratize heritage access and encourage community-based heritage stewardship.

Keywords Heritage · GIS · Educommunication · Cybercommunities · Web 2.0

Introduction

Since 2016, the University of Zaragoza has developed various applications based on Geographic Information Systems (GIS) that allow the collaborative cataloguing of cultural heritage. These tools are freely available for use on computers, tablets and mobile phones.

All these applications are built on a shared foundation comprising two key elements: the geolocation of heritage assets on a map using GIS and the textual or

P. Rivero (✉) · S. García-Ceballos · B. Aso
ARGOS-IUCA Research Group, Faculty of Education, University of Zaragoza, Zaragoza, Spain
e-mail: privero@unizar.es

I. Navarro-Neri
Research Group Life Long-Learning (UPNA I-COMMUNITAS), Universidad Pública de Navarra, Pamplona, Spain

R. de Miguel González and P. Rivero Gracia (eds.), *Geospatial Technologies for Heritage Education*, Key Challenges in Geography,
https://doi.org/10.1007/978-3-031-96372-8_11

visual contributions that users provide to the assets through images, comments and reflections they wish to share, aimed at digitally preserving and safeguarding them. These contributions are shared with the aim of digitally preserving and safeguarding the assets within the app.

The research group leverages this approach to design and implement various citizen science initiatives within digital environments. In these spaces, users engage in participatory processes by sharing, co-creating or exchanging their own images, videos, and/or audio recordings of different heritage assets, starting from a specific physical location transferred to a GIS-based application. These assets are then gathered in a digital repository that captures the cultural richness of different places and communities. Users can also engage with each other by commenting on posts and sharing content across social media. These spaces are built on an open horizontal participatory model that fosters learning by overcoming geographical barriers. They provide an active process of ongoing construction, reinterpretation, resignification and reflection, creating a shared and geolocated culture that facilitates the creation of knowledge nodes that transcend physical space.

This citizen science project primarily rests on three main theoretical foundations: (1) the geolocation and digitisation of heritage to ensure its preservation, collaborative cataloguing and dissemination that lead to the formation of heritage cybercommunities; (2) the 'R-elational factor' (Marta-Lazo and Gabelas 2016), which helps us to understand the evolution from ICTs to RICTs; and 3) the design of educational proposals for both formal and non-formal learning environments, aimed at activating the co-creative paradigm.

This chapter provides detailed information on the collaborative applications hosted on the website https://civitas.unizar.es: *Mural Hunter, Arqueovirtual* and *CatPat*. In addition, in the section dedicated to each application, various educational proposals are developed for implementation in formal educational settings, such as secondary schools and universities.

Theoretical Framework

The Geolocation and Digitisation of Heritage: Cataloguing and Dissemination to Foster the Emergence of Heritage Cybercommunities Through the Use of Apps

Since their inception in the mid-twentieth century, GIS have been defined in several ways according to the various approaches through which they can be examined (Arcila 2003). In essence, they are computer packages designed to store, visualise, manipulate and analyse geographically referenced data (Seguí et al. 2012). When applied to our field, georeferencing holds significant educational potential (De Miguel 2016), as knowing the precise location of a heritage asset facilitates contextualised learning about its history or culture. This is achieved by providing users with detailed data, information and/or multimedia resources about the site, thereby deepening their understanding. Moreover, the geolocation of heritage can spark curiosity about the importance of these assets to the local community, encouraging users to explore their place of origin. These tools not only preserve cultural memory, but they also transform how people relate to their heritage.

The democratisation of access to technology and technical advancements at all levels over the past 25 years has revolutionised how we approach heritage. In this context, various educational and social agents such as schools, universities and museums have adapted to this new, unstable and ever-evolving landscape (Ibáñez-Etxeberria et al. 2019).

The digital preservation of heritage has now become a consolidated trend thanks to initiatives arising from a wide range of organisations, including museum networks, libraries and universities. These actions not only focus on cataloguing heritage assets, but they also aim to democratise access to heritage (Rivero et al. 2020a, b).

These initiatives promote new models of social interaction and connection, such as networked individualism (Castells 2009) and hypermediations (Scolari 2008), leading to the concept of cybercitizenship as defined by Badillo Mendoza and Marta-Lazo (2019):

> [...] the various relations or manifestations of the subjects that underpin a society, exercise rights and duties and, depending on certain situations, mobilise collective actions from an ecological vision of communication, acting from a reasoning that integrates cybercultural and situated politics. (p. 147)

Cybercitizenship aims not only to stay informed or mobilised through the net or social media, but also to promote dialogue that transcends physical and mental barriers often encountered in the digital sphere (Navarro-Neri et al. 2021a). Within this digital context, heritage education understands that this environment opens the way to democratisation, socialisation and education in and with heritage (Fontal 2006), thereby introducing the concept of heritage educommunication.

Previous research (Rivero et al. 2020a, b) has demonstrated that educommunicative processes within social web contexts can effectively foster true heritage cybercommunities when participants not only understand the value of that heritage, but also establish a specific connection to it through territorial identification via geopositioning. Furthermore, as highlighted by other studies (Sebastián-López and de Miguel González 2020), the use of GIS favours the development of the comprehensive TPACK (Technological-Pedagogical Content Knowledge) model, which in this case allows teachers to integrate digital competences alongside those specific to geography and history through participatory, collaborative and experiential processes (de Miguel et al. 2024).

In essence, the shared identity that emerges from collaborative networked actions fosters the formation of heritage cybercommunities, strengthened by emotional connections to heritage (Rivero et al. 2018, 2020a, b; Santacana i Mestre and Martínez Gil 2018). This, in turn, facilitates the development of educational proposals such as those presented in this chapter in order to enhance educommunicative processes through consolidated heritage cybercommunities that can be created in real time, while visiting the places or assets, thereby enriching the experience by connecting theory with practice, while also allowing for the interactive visualisation and analysis of spatial data.

The 'R-Elational Factor' and the Transition from ICTs to RICTs

The 'R-elational factor' is a term coined by Marta-Lazo and Gabelas to define 'the axis connecting the uses, consumptions and interactions of participants in the digital sphere' (Marta-Lazo and Gabelas 2016, p. 31), marking the transition from ICTs (Information and Communication Technologies) to RICTs (Relationship, Information and Communication Technologies).

For these authors, the development of the digital competence of the *emirec*—an individual who is both a sender and a receiver of content in the digital sphere (Cloutier 1973)—will help us to function in a responsible, ethical and supportive manner in the future (Marta-Lazo and Gabelas 2016, p. 120). This perspective follows a holistic and positive approach that encompasses the cognitive, social and emotional dimensions of the human being and the social agents and mediators in individual, social and environmental settings (Marta-Lazo and Gabelas 2016, p. 93).

The 'R-elational factor' takes shape in what its authors call 'inteRmethodology': a proposal that seeks convergence between different methodological options through the use of various active techniques that help students to acquire competences of all kinds, whether cognitive or based on participation (Marta-Lazo and Gabelas 2016). This fosters spaces that favour individual exploration, participation, critical analysis and content creation.

All this occurs within what is described as a 'fourth environment': a hybrid and symbiotic ecosystem that seamlessly merges offline reality with the multi-screen online world, accessible anytime and anywhere (Gabelas 2015). This environment is characterised by high connectivity, mobility, portability and ubiquity, where interactions become more flexible and communications move to a space of both technological and media convergence (Villalonga Gómez and Marta-Lazo 2016).

Thus, the R-elational factor provides an opportunity to review and update the discourse on the use of ICTs that can 'break' with the trend by reviewing and analysing the current context from conceptual and cultural frameworks typical of the past (Hergueta-Covacho et al. 2016).

The evolution from ICTs to RICTs is framed within dialogical educommunication,[1] since 'technology is a means for relationship and participation, rather than an end in itself' (Marta-Lazo et al. 2016, p. 308). The concept of RICTs is directly related

[1] Dialogical educommunication and media literacy are terms based on the ideas of Freire (1973, 2005) and Kaplún (1998) and synthesised by Aparici (2010) as 'a philosophy and practice of education and communication based on dialogue and participation, requiring not only require technologies but also a change of attitudes and conceptions' (p. 12). This definition integrates, on one hand, Paulo Freire's concept of dialogic education (1973, 2005), which stresses that education is a dialogical

to media convergence (Gabelas et al. 2015) and allows us to abandon elitist conceptions in formal and non-formal education as well as to produce teaching processes and spaces based on collaborative, social and cultural practices (Marta-Lazo et al. 2016).

This transition from ICTs to RICTs through the R-elational factor manifests itself in how it transforms the user's relationship with each element of ICTs. Firstly, it transforms the relationship with technology, allowing users to transfer how they express themselves in their online leisure time into learning spaces thanks to a new hypertextual language that turns users into consumers, creators, redefiners and ambassadors of content, thereby promoting interaction within digital competences. Secondly, it transforms the relationship with information by encouraging Internet users to analyse and develop critical thinking while searching for content; this not only promotes intellectual maturity, but also enhances synaptic capacity, as users must establish connections to contextualise information. Finally, it transforms communication, as promoting the social connections that allow the exchange of ideas makes it possible to implement the feed-feed model (Aparici and Silva 2012) to build knowledge through the use of mobile devices (Gabelas et al. 2015; Gabelas and Marta-Lazo 2020).

In summary, the 'R-elational factor' that underpins the evolution from ICTs to RICTs provides an opportunity to create spaces that support learning and foster the appreciation of heritage, thereby promoting synergies among individuals with common cultural interests.

The Co-creative Paradigm for Designing Educational Proposals in Spaces of Geographic, Historical and Heritage Education

GIS plays a fundamental role in geography by generating, processing and representing geographic information for educational purposes beyond merely locating a specific element in the territory (Nieto 2016). As Marta-Lazo (2018) points out, leveraging this consolidated technological development allows us to move from an educommunication based on the paradigm of 'transmission' to one centred on 'mediation'. In other words, this involves reappropriating knowledge by engaging in rational reflection on interpretations, thereby emphasising the process and the individuals (Rivero et al. 2021).

GIS and geolocation are key tools in heritage applications, as they allow for more effective and enriching heritage management. Among other functionalities, they help users to visually and comprehensively represent the locations of heritage

process in which meanings are woven and identities and a sense of belonging are constructed as a foundation for transforming reality (Rodríguez García 2017). On the other hand, it incorporates Kaplún's model of pedagogical communication (1998), which asserts that 'every human being is endowed and empowered for both functions and has the right to participate in the communication process, alternately acting as sender and receiver' (p. 65).

sites, thus enhancing spatial understanding of their distribution and connection to the environment. Geolocation in mobile apps assists users in easily locating and accessing sites of heritage interest. Using these apps, therefore, serves not only to document heritage, but also to make it known and encourage visits and knowledge transfer. Any observation made by users regarding the condition of these assets can provide crucial information for the conservation and sustainable use of heritage. In addition, in terms of educational profitability for institutions or research, GIS offers the potential for spatial analysis that can uncover patterns and trends in visits to heritage sites, aiding in the planning of activities and the allocation of resources. These systems also help in the analysis of geolocation data. Moreover, one of the most important aspects for us is that applications that employ geolocation may include functions that users can use to contribute information or record their experiences, promoting greater participation with heritage and awakening a sense of community participation, appropriation or belonging and thereby encouraging citizen science initiatives.

This chapter proposes the apps *Mural Hunter*, *CatPat* and *Arqueocivitas* as meeting places. These are spaces for mediation and the establishment of interactive and rhizomatic communicative processes,[2] emulating successful dynamics implemented by museum institutions in the Web 2.0 (Aso 2021; Rivero et al. 2021; Sebastián Novell 2019). All these processes are horizontal and promote rhizomatic learning (Cormier 2008), as rhizomes grow and spread in a 'nomadic' manner, subject only to the constraints of their own habitat (Cabero Almenara and Llorente 2015, p. 190).

Therefore, based on the definition of the Museum 2.0 by Rivero et al. (2021, p. 16), the apps presented in this chapter serve as spaces for encounter and common creation that are revealed as reticular, heterotopic, panoptic and enabling of multidirectional participation that encourages co-creation, which is the main focus of the relationship between users, with the University of Zaragoza acting as the manager of these apps.

Through this commitment to horizontal and rhizomatic educommunicative spaces, we can frame the didactic proposals presented in this chapter within interactive pedagogy (Aparici and Silva 2012), which opposes transmissive pedagogy and supports dialogue and constructive co-authorship of knowledge (Rivero and Aso 2023). This is commitment to citizen science mediated through digital devices or e-Citizen Science (Finquelievich and Fischnaller 2014) is aimed at engaging various educational agents (universities, institutes, museums, etc.) in civic and social decision-making concerning heritage, such as efforts to catalogue it or use it as an element of dialogue and debate.

---

[2] Deleuze and Guattari (1985) introduced the theory of rhizomatic thinking, using the rhizome as a metaphor for a new mode of horizontal thinking that contrasts with traditional, vertical or hierarchical models of thought.

Apps and Educational Approaches

As mentioned at the beginning of this chapter, since 2016, the University of Zaragoza has been developing apps for the collaborative cataloguing of cultural heritage based on GIS. Available on the *civitas* website (https://civitas.unizar.es), these apps serve as both citizen science projects and approaches to heritage education in the non-formal sector. They are now being implemented in formal education settings, specifically in secondary schools and universities.

All the apps share a common foundation. Users can upload images or videos of heritage assets onto a map accompanied by basic data and a brief description or comment, which can subsequently be shared on social media platforms. The applications are optimised for mobile devices across various operating systems, but they can also be accessed via desktop computers.

In selecting a GIS platform, more comprehensive options with enhanced educational potential were dismissed due to their inability to guarantee immediate technical assistance and supervision during software use, as is the case with ArcGIS. Consequently, Google's system was chosen as the foundational platform for two key reasons. Firstly, it is a free geolocation system that requires minimal financial investment for maintenance, ensuring the long-term sustainability of the apps, and its updates are compatible with the vast majority of mobile devices available on the market. Secondly, its user interface is straightforward, intuitive and familiar to most users, particularly due to its widespread use in GPS technology. This familiarity helps in implementing educational approaches with no need for technical training in geopositioning, as users can engage with the tool intuitively.

Mural Hunter *(Mural)*

The *Mural Hunter* app (called 'Mural' in its web version) is designed for cataloguing, digitally preserving and interpreting urban or mural art, as well as for dialoguing with it (Rivero and Gil-Díaz 2016). This app was developed as part of the project EDU2016-78163-R: HERITAGE EDUCOMMUNICATION WEB 2.0, MINECO FEDER (Retos Call), active from 30 December 2016 to 29 December 2019.

This project was established with two primary objectives: firstly, to create a global catalogue of wall art; and secondly, to do so through a collaborative citizen science initiative encouraging a high degree of creative interactivity. In doing so, the project aims to foster the emergence of a heritage cybercommunity, united by transnational urban mural art and comprised of an active citizenry dedicated to protecting heritage, which would in turn contribute to the patrimonialisation of urban mural art (Rivero et al. 2020a, b).

The project was designed as a citizen science initiative in the sense that users themselves, without any prior specialised training, geoposition their photographs of urban art murals in the application using their computers or mobile devices.

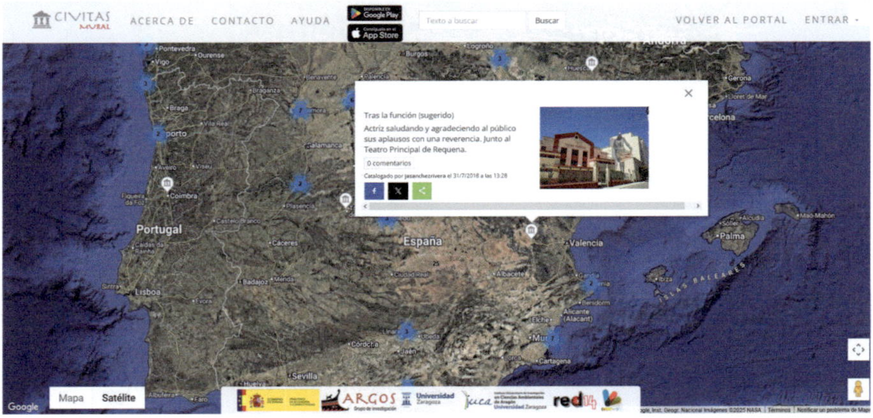

Fig. 1 Screenshot of the app "Mural Hunter" (Courtesy of https://civitas.unizar.es/mural)

Users of this application can upload images that capture various expressions, creations or manifestations of urban art from around the world. After adding the geolocation, users attach the corresponding image or images and can provide additional information, such as a title and the name of the creator if they know it. Most interestingly, there is a section for comments where users can express their reasons for sharing the mural and communicate their impressions (their feelings upon viewing the artwork, any associations it may evoke with other works or places, any concerns regarding its theme or condition or any other appreciations) (Fig. 1).

Thanks to this dynamic approach, the app can be seen as a tool designed to build bridges between people from all over the world—regardless of their physical location—who share a common interest in promoting the value of urban heritage. It also serves as a powerful photographic archive for ephemeral artistic expressions, preserving the identity of the works themselves. Acting as a documentary repository for urban art, it is entirely driven by citizen participation or through educational initiatives mediated by citizen science.

Additionally, thanks to the possibility of uploading images of the same artwork, users can easily track its progression, from its creation to its possible disappearance or replacement by a new piece in the same space, as the case may be. This ensures that artworks that have disappeared from the physical world continue to exist in the digital realm (Navarro-Neri 2021).

The app currently hosts a catalogue of over 1300 geopositioned artworks and has more than 300 registered users. It also has its own Facebook page with more than 200 followers.

Educational Proposal

Based on citizen co-creation, these applications also include educational approaches linked to their use that are sometimes implemented in both formal and non-formal settings for their study and development. All the applications are grounded in the three theoretical pillars outlined in this chapter, supported by the accompanying bibliography. Short presentations have been prepared in each case highlighting the key aspects of these theories to facilitate and streamline the teaching and learning process with teachers and students, as shown below.

Prior to engaging in the activities, it is advisable to conduct a preliminary questionnaire with the students. This questionnaire should ask them a range of questions, such as whether they know what mural art is, whether they can provide their own definition of it and whether they have ever seen examples of it and where. Additional questions may also be included, such as whether mural art should be preserved, whether it should be studied and analysed (and if so, how this should be done, as is the case with sculptures, paintings and buildings in specific artistic styles) and whether it must necessarily be subversive or if it can be aligned with the official narratives of those commissioning the work.

To conduct an activity with secondary school students (aged 14 to 18), we believe it is particularly interesting to begin with their immediate surroundings: the route they take from home to school, their neighbourhood and their town or village. If the activity takes place in small rural communities, we can use examples shared by the artists themselves on social media, although these artistic expressions are becoming increasingly prevalent even in rural locations to revitalise, decorate, enhance or modernise them.

Activity before using *Mural Hunter*: In class, students will explore the concepts of urban mural art and graffiti to introduce the type of heritage to be catalogued. In addition, topics focused on developing the digital competences that all students should acquire will be addressed. These include researching information related to mural art (e.g., the origin and creator of the piece), geopositioning using GIS and building a hypertext to describe the artwork to be shared on the app. The application will also be introduced and students will have the opportunity to browse it to see which murals have already been uploaded.

Activity while using *Mural Hunter:* This educational proposal suggests that each student uploads three examples of mural art to the application, following all the requirements outlined in it: the most precise location possible, the name of the piece if they know it (or they may suggest one) and the name of the creator, if they know that. In the comment section, students will need to construct a hypertext that includes an objective description of the mural, an analysis of the theme depicted in the artwork and their personal opinion, where relevant. The text should also include hyperlinks to supplement the information and a hashtag to make it easier to share the content on social media.

If the activity is conducted with university students, such as future teachers or fine arts students, we can apply the same approach used for secondary school students. To add complexity, students could be asked to upload a photo they have taken during a trip to *Mural Hunter* and explain in the comments section why they took the photo and whether they would have done so if the mural were part of their daily surroundings, such as if they saw it every day on their way to university.

In essence, this educational proposal using *Mural Hunter* allows students to: (1) develop their digital competence in areas such as research, the use of GIS and hypertext creation; (2) work with both theoretical and practical aspects of concepts such as educommunication of heritage; and 3) actively participate in a collaborative activity that is characteristic of citizen science. For all these reasons, we believe that if the activity is to be assessed, the criteria should focus on the development of these three specific competences.

CatPat *(*CatPat*)*

Available in both a web version at https://civitas.unizar.es/catpat and as an app for Android and iOS mobile devices, the *CatPat* app is a newly created tool launched by the ARGOS research group for public use. It was recently promoted in educational contexts through the project TED2021-131174B-100 *Participatory Digital Tools for Supporting Cyberheritage Communities* (AEI).

CatPat undertakes a collaborative mapping of tangible and intangible public heritage in which users are asked to provide information and personal opinions, similarly to the *Mural Hunter* application. Users are prompted to input details such as the title of the heritage asset, its geopositioning and a free-text commentary where they can include additional information, opinions and hyperlinks for further exploration, etc.

This app introduces two new features not found in *Mural Hunter*. Firstly, it allows users to include audio notes so they can contribute in this format or document intangible heritage like popular songs or local stories that they deem important for others to know. Secondly, all uploaded and geopositioned heritage items must answer the question of whether they are considered endangered. The app also includes visual cues, with circular icons representing tangible heritage and diamond-shaped icons for intangible heritage. Icons are white for non-endangered heritage and red for endangered heritage (Fig. 2).

Therefore, *CatPat* should not be classified as a tool for creating or enriching existing digital collections, but rather as a way to facilitate interactions and social connections (García-Ceballos and Rubio-Navarro 2023). From a 'R-elational' perspective, this collaborative app not only documents and adds value to users' contributions, but it also fosters the emergence of new heritage concepts.

As mentioned above, the application is currently being promoted in various schools and universities through the project TED2021-131174B-100 *Participatory Digital Tools for Supporting Cyberheritage Communities* (AEI). *CatPat* plays a crucial role in the MOOC *Cultural Heritage in the RICT Environment*. After learning about concepts such as educommunication and RICTs, the R-elational factor and

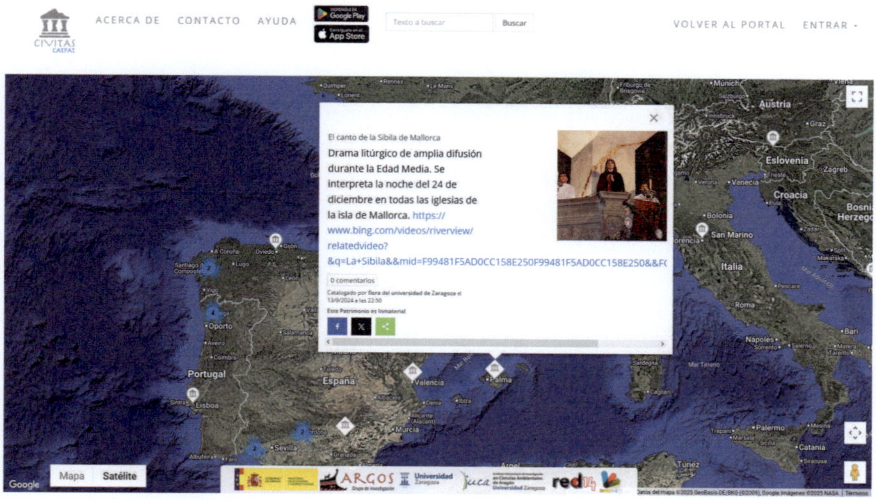

Fig. 2 Screenshot of the app "CatPat" (Courtesy of https://civitas.unizar.es/catpat)

public art as heritage, participants in this free course must learn to use the *CatPat* application and upload several contributions.

The application already hosts 300 contributions and has more than 100 registered users. The number of contributions and registered users is steadily increasing, as educational activities are continuously being implemented in schools and universities. Below is a summary of an educational proposal that has already been successfully implemented.

Educational Proposal

Like the educational proposal developed for the *Mural Hunter* app, this one begins with theoretical presentations on key concepts such as the definition and various types of heritage (Cuenca 2002; Fontal 2003), as well as reflections on when a heritage asset should be considered at risk. References to heritage educommunication and the development of digital competence have also been included.

In this case, it is also useful to conduct a preliminary questionnaire where students are asked to define heritage in their own terms, name five examples of tangible heritage and five examples of intangible heritage without searching online, provide examples of endangered heritage and consider anonymously whether they have ever endangered a heritage asset themselves.

Since the activity involves the collaborative mapping of both tangible and intangible heritage, students are recommended to focus on local, regional or national heritage assets. In this way, the arguments they make for selecting a specific heritage asset will offer insights into issues of identity, for example.

Activity before using *CatPat*: The students will explore the theoretical framework summarised in a presentation addressing key topics like the definition of heritage, its various forms, examples of heritage at risk and the threats that heritage has faced and continues to face. The activity will also cover proper geopositioning using GIS, the application's functionalities and strategies for building a well-constructed hypertext rich in content and additional resources. Finally, the students will be introduced to the *CatPat* app and encouraged to browse through existing contributions.

Activity while using *CatPat*: Each student will be asked to contribute four heritage items to the app—two tangible and two intangible assets. As mentioned above, the activity should be limited to the local, regional or national context. Additionally, students will be tasked with leaving at least five comments on other contributions to promote dialogue around the heritage items.

A roundtable discussion is recommended to conclude the activity. During this discussion, students will share the reasons behind their heritage asset selections and debate key issues, such as whether the chosen assets deserve preservation, why, and whether these assets are part of a collective memory or simply personal choices.

In short, this approach allows students to engage in an activity starting with the geolocation of a heritage asset through GIS and encourages reflection on the value of intangible heritage. It also introduces them to the dynamics of collaborative citizen science and enables exploration of the symbolic and identity-related connections that exist or could emerge with a heritage asset (García-Ceballos and Rubio-Navarro 2023).

RomanSites (ArchaeoVirtual)

The *RomanSites* application (called *Arqueovirtual* in its web version, available at https://civitas.unizar.es) is an open online collaborative space whose main objective is to use geopositioning to catalogue digitalisations and reconstructions of Roman archaeological sites and maintain a registry of the various social media profiles that display, inform and disseminate content about these sites (Fig. 3).

Conceived as a citizen science initiative, this project has been developed by the University of Zaragoza like the aforementioned applications and was funded through the research project HAR2013-48,456-C3-3-P: 'Qualitative evaluation of the cultural action of Roman archaeological sites with Web 2.0 (CIVITAS)'.

Like *Mural Hunter* and *CatPat*, the *RomanSites* app is based on a GIS system where users share information on digitisations and reconstructions of Roman archaeological sites after geolocating them on the map provided. Users must then enter the name of the site and indicate whether it is a digitisation, a virtual reconstruction or a social media profile. Finally, the application invites users to leave a description of the site and the website of the place to be shared in an annexed box.

In this way, through the cooperative construction of knowledge, resources are provided to implement heritage education actions in formal educational contexts

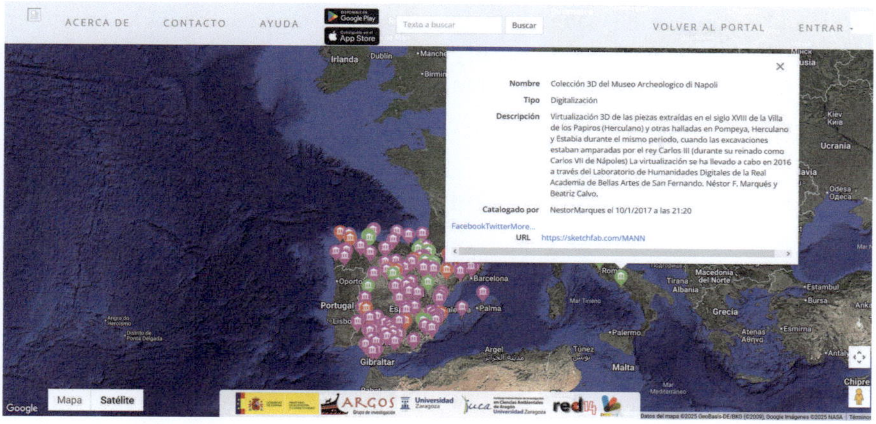

Fig. 3 Screenshot of the app "Arqueovirtual" (Courtesy of https://civitas.unizar.es/arqueovirtual)

such as virtual reconstructions of geopositioned Roman sites, social media profiles and/or websites with complementary material (Rivero and Gil-Díez 2016).

Educational Proposal

In this case, we must bear in mind that we are dealing with a more specific type of heritage—archaeological heritage—that pertains specifically to Greco-Roman culture. Therefore, to use this geopositioning tool effectively, the teachers involved should receive prior training in a robust epistemological framework. This framework should encompass not only the history of Ancient Rome, but also key concepts related to archaeology, digital reconstructions and archaeological museums and their use of social media.

The research suggests a didactic sequence based on investigating a Roman archaeological site across several dimensions. Firstly, students will explore the history of the site using available documentary sources and material evidence. Next, they will turn to the digital sphere to gather information, such as current news about the site, the status of ongoing excavations or research, the availability of digital reconstructions in video or virtual reality formats and whether any museums or curated exhibition spaces with an official social media presence exist for the site.

With all this information, students will compile their findings into two outputs, either individually or in groups: 1) an entry in the application by filling in the required fields (name, description, website, etc.); and 2) a portfolio and presentation of the collected information, with the possibility of approaching the content longitudinally or by thematic interest.

Due to the complexity of this topic, this research suggests that it is feasible to implement this activity with students aged 12 and 13, provided the teacher carefully selects and supervises the materials to be analysed and studied. Similarly, for university settings, it is recommended to carry it out with undergraduate history students or those enrolled in a master's programme in education.

Adaptation to an Analogue Context

As can be observed, all the educational proposals presented in this chapter have been designed in a context that ensures access to digital media or devices, assuming that all students have independent access to a computer, tablet or mobile phone. However, if these educational activities are implemented in schools with a digital divide, the research group will provide the necessary devices to facilitate the activity, with the support of its members.

Alternatively, if one of the proposals is to be implemented but not all students have access to the required technology, the teacher can be responsible for uploading the materials and comments to the relevant application. In this way, students can continue to develop their geographical thinking through the use of analogue methods while simultaneously working on heritage educommunication and digital competence.

Conclusions

Accustomed to the transmissive nature of heritage educommunication by museums, these proposals herald a paradigm shift, as the institution managing the technological resource—in this case the University of Zaragoza—not only shares a list of geopositioned heritage assets, but also provides a digital environment and invites participation in it. This paradigm shift allows us to move from mere transmission to mediation and to the co-creation of new meanings around heritage.

It is also important to highlight the relevance of these activities. For example, in the case of *Mural Hunter*, mural art is clearly already recognised and accepted as an artistic discipline. However, due to its location in public spaces or perhaps a lack of sensitivity and education towards this form of art, many works are strikingly often vandalised or destroyed soon after their completion.

It is true that the critical or controversial and often ephemeral nature of this type of heritage can make it difficult for the inhabitants of the place where it is located to feel like it is their own or is part of their community identity. However, this application and its educational proposal provide an opportunity for dialogue about these and other aforementioned issues.

The value of *Arqueocivitas* lies primarily in how it serves as a new resource bank for students and teachers at all levels, providing access to key Web 2.0 spaces that deal with Roman archaeological sites (museums, sites, etc.) and 3D reconstructions through programs like Sketchfab. This opens new avenues of research in education, such as the use of virtual reality, and expands on existing ones, like the employment of virtual reconstructions as teaching tools.

Finally, the current and future importance of the *CatPat* application must be emphasised. By using it in educational settings, teachers and students can engage in participatory and co-creative dynamics where heritage becomes the axis and driving force of new identity processes. These processes directly boost the appreciation, protection and dissemination of local heritage.

Acknowledgements This work is supported by the project 'Digital Competences, Learning Processes, and Cultural Heritage Awareness: Quality Education for Sustainable Cities and Communities' (code PID2020-115288RB-I00), funded by the State Research Agency (AEI), Ministry of Science and Innovation of the Government of the Kingdom of Spain (AEI/10.13039/501100011033).

References

Aparici R, Silva M (2012) Pedagogía de la interactividad. Comunicar 38:51–58. https://doi.org/10.3916/C38-2012-02-05

Aparici R (Coord) (2010) Educomunicación: más allá del 2.0. Gedisa

Arcila M (2003) Sistemas de información geográfica y medio ambiente: principios básicos. Servicio de Publicaciones de la Universidad de Cádiz

Aso B (2021) Educación patrimonial en tiempos digitales. Estudio de la educomunicación en redes sociales del Museo Diocesano de Jaca [Tesis Doctoral, Universidad de Zaragoza]

Badillo Mendoza ME, Marta-Lazo C (2019) Ciberciudadanía a través de Twitter: caso Gran Marcha Carnaval y consultas populares contra la minería en La Colosa. Cuadernos.Info 45:145–162. https://doi.org/10.7764/cdi.45.1454

Cabero Almenara J, Llorente MDC (2015) Tecnologías de la Información y la Comunicación (TIC): escenarios formativos y teorías del aprendizaje. Revista Lasallista de Investigación 12(2):186–193. https://www.redalyc.org/articulo.oa?id=69542291019

Castells M (2009) Comunicación y Poder. Alianza

Cloutier J (1973) La communication audio-scripto-visuelle à l'heure des self média. Les Presses de l'Université de Montreal

Cormier D (2008) Rhizomatic education: Community as curriculum. Innovate: Journal of Online Education 4(5):2

Cuenca JM (2002) El patrimonio en la didáctica de las ciencias sociales: análisis de concepciones, dificultades y obstáculos para su integración en la enseñanza obligatoria [Tesis Doctoral, Universidad de Huelva]. http://rabida.uhu.es/dspace/handle/10272/2648

De Miguel R (2016) Del pensamiento espacial al conocimiento geográfico a través del aprendizaje activo con tecnologías de la información geográfica. Giramundo: Revista de Geografia do Colégio Pedro II 2(7):7–13. https://doi.org/10.33025/grgcp2.v2i4.668

De Miguel R, Rivero P, Kratochvil O (2024) Competencias digitales, históricas y geográficas y de comunicación patrimonial: los storymaps como recursos de aprendizaje en la formación del profesorado de ciencias sociales. Andamio. Revista en Didáctica en Historia 6(1):108–129. https://revista-andamio.cl/index.php/revista/article/view/141/119

Deleuze G, Guattari F (1985) El anti Edipo. Capitalismo y esquizofrenia. Paidós

Finquelievich S, Fischnaller C (2014) Ciencia ciudadana en la Sociedad de la Información: nuevas tendencias a nivel mundial. Revista Iberoamericana de Ciencia, Tecnología y Sociedad 9(27):11–31. Disponible en: http://www.scielo.org.ar/scielo.php?script=sci_arttext&pid=S1850-00132014000300002&lng=es&tlng=es

Fontal O (2003) La educación patrimonial: definición de un Modelo Integral y diseño de sensibilización [Tesis Doctoral, Universidad de Oviedo]

Fontal O (2006) Las webs: complementos y extensiones de los museos. En Calaf R, Fontal O (Coords.) Miradas al patrimonio. Trea, pp 181–200

Freire P (1973) ¿Extensión o comunicación? La concientización en el medio rural. Siglo XXI

Freire P (2005) Pedagogía del oprimido. Siglo XXI

Gabelas JA (2015) Cuarto entorno. INED 21. Recuperado de: http://ined21.com/cuarto-entorno/

Gabelas JA, Marta-Lazo C (2020) La era TRIC: Factor R-elacional y educomunicación. Ediciones Egregius

Gabelas JA, Marta-Lazo C, González-Aldea P (2015) El factor relacional en la convergencia mediática: una propuesta emergente. Anàlisi. Quaderns de Comunicació i Cultura 53:20-34. https://doi.org/10.7238/a.v0i53.2509

García-Ceballos S, Rubio-Navarro A. (2023) ¿Cuántos somos necesarios? CATPAT, aplicación colaborativa para la creación de cibercomunidades patrimoniales en contextos escolares. Revista PH 109:166–168. https://doi.org/10.33349/2023.109

Hergueta-Covacho E, Marta-Lazo C, Gabelas-Barroso JA (2016) Educación Mediática e InteR-metodología Relacional aplicada a los MOOC. Revista Mediterránea de Comunicación 7(2):47–58. https://doi.org/10.14198/MEDCOM2016.7.2.4

Ibáñez-Etxeberria A, Kortabitarte A, De Castro P, Gillate I (2019) Competencia digital mediante apps de temática patrimonial en el marco DigComp. Revista Electrónica Interuniversitaria de Formación del profesorado 22(1):13-27. https://doi.org/10.6018/reifop.22.1.356231

Kaplún M (1998) Una pedagogía de la comunicación. De la Torre

Marta-Lazo C (2018) La brújula de la educomunicación [Editorial]. Revista Mediterránea De Comunicación 9(2):7–8. https://doi.org/10.14198/MEDCOM2018.9.2.24

Marta-Lazo C, Gabelas JA (2016) Comunicación digital. Un modelo basado en el Factor R-elacional. Editorial UOC

Marta-Lazo C, Marfil-Carmona R, Hergueta-Covacho E (2016) Aplicación de las Redes Sociales en el aprendizaje en conectividad: uso del factor relacional en la dialéctica *Twitter*. Revista científica electrónica de Educación y Comunicación en la Sociedad del Conocimiento 16(2):304–319. http://eticanet.org/revista/index.php/eticanet/article/view/108

Navarro-Neri I (2021) Mural Hunter: una app de catalogación y diálogo en torno al patrimonio artístico urbano. Revista PH 103:124–126. https://doi.org/10.33349/2021.103.4871

Navarro-Neri I, Aso B, Mendioroz-Lacambra A (2021a) Análisis del lenguaje educomunicativo de los museos arqueológicos españoles durante el confinamiento de 2020. Clio. History and History Teaching 47:132–160. https://doi.org/10.26754/ojs_clio/clio.2021475522

Navarro-Neri I, Rivero P, Aso B, Mendióroz-Lacambra A (2021b) Narrativa y significación histórica en la comunicación de los museos nacionales canadienses en Facebook. Panta Rei. Revista digital de Historia y Didáctica de la Historia 15:185–207. https://doi.org/10.6018/pantarei.467881

Nieto G (2016) Análisis de la práctica educativa con SIG en la enseñanza de la geografía de la educación secundaria. Un estudio de caso en Baden-Wüttemberg, Alemania [Tesis doctoral, Universidad de Barcelona]. http://hdl.handle.net/10803/400097

Rivero P, Aso B (2023) Ciencia ciudadana mediada por dispositivos digitales: desafíos y reflexiones en torno al patrimonio cultural. Revista PH 109:172–174. https://doi.org/10.33349/2023.109

Rivero P, Gil-Diez I (2016) Ciencia ciudadana, arqueología y educación patrimonial. En: López Facal R (ed) Libro de Actas del Simposio Ciencias sociales, educación y futuro. Investigaciones en didáctica de las ciencias sociales. Santiago de Compostela: Universidad de Santiago de Compostela, pp 578–585

Rivero P, Fontal O, García-Ceballos S, Martínez-Rodríguez M (2018) Heritage Education in The Archaeological Sites. An Identity Approach in The Museum of Calatayud. Curator 61(2):315–326. https://doi.org/10.1111/cura.12258

Rivero P, Navarro-Neri I, Aso B (2020a) Educommunication Web 2.0 for Heritage: A View from Spanish Museums. En Delgado-Algarra EJ. Cuenca-López JM (eds) Handbook of research on citizenship and heritage education, pp 449–470. https://doi.org/10.4018/978-1-7998-1978-3.ch021

Rivero P, Navarro-Neri I, García-Ceballos S, Aso B (2020b) Spanish archaeological museums during COVID-19 (2020): an edu-communicative analysis of their activity on Twitter through the sustainable development goals. Sustainability 12:8224. https://doi.org/10.3390/su12198224

Rivero P, Jové G, Sebastian C (2021) Educomunicación en las redes sociales de los museos en la era post covid: el paradigma co-creativo. Her & Mus. Heritage & Museography 22:8–17. https://doi.org/10.34810/hermusv22id394966

Rodríguez García YM (2017) Reconceptualización de la educación en la era digital. Educomunicación, redes de aprendizaje y cerebro: una visión desde la neurociencia cognitiva a los procesos de construcción de conocimientos en entornos digitales [Tesis Doctoral, Universidad Nacional de Educación a Distancia]

Santacana i Mestre J, Martínez Gil T (2018) El patrimonio cultural y el sistema emocional: un estado de la cuestión desde la didáctica. Arbor 194(788):a446. https://doi.org/10.3989/arbor.2018.788n2006

Scolari C (2008) Hipermediaciones. Elementos para una teoría de la comunicación digital interactiva. Ed. Gedisa

Sebastián Novell C (2019) La historia de vida como herramienta para la evaluación de los espacios 2.0 de los museos [Tesis Doctoral, Universitat de Lleida]

Sebastián-López M, de Miguel González R (2020) Mobile learning for sustainable development and environmental teacher education. Sustainability 12:9757. https://doi.org/10.3390/su12229757

Seguí AE, Portalés C, Cabrelles M, Lerma JL (2012) Los sistemas de información geográfica: concepto, ventajas y posibilidades en el campo de la restauración. Loggia, Arquitectura & Restauración 24–25:122–131. https://doi.org/10.4995/loggia.2012.3008

Villalonga Gómez C, Marta-Lazo C (2016) Factor relacional y colaboración con “Apps” en la asignatura Metodología de la Investigación. F@ro: revista teórica del Departamento de Ciencias de la Comunicación 1(23):195–212. https://www.revistafaro.cl/index.php/Faro/article/view/466/443

Geographic Information Systems and Teacher Training in Heritage Education

Andrés Domínguez-Almansa, Tania Riveiro-Rodríguez, Ramón Blanco-Chao, and Manuela Costa-Casais

Abstract This work posits that the study of space, territory, and the various locations subject to processes of heritage appropriation should be part of the initial teacher training. It advocates for the use of Geographic Information Systems (GIS), which enable the incorporation of new innovative elements in the teaching of social sciences. GIS no longer requires complex technical training, so pre-service teachers can learn to use them without great difficulty. Therefore, two complementary options are proposed: training teachers in the necessary skills to use software and data, and training them to develop didactic approaches that integrate technical learning and environmental research. From this perspective, three fundamental questions are raised: (i) Technical learning can and should evolve on various levels; (ii) Technical training can only be considered (in combination) with knowledge of the subject being analysed; and (iii) The learning teachers acquire in training will be transferred to new educational scenarios, using GIS as a reference.

Keywords Geographic information systems · Natural heritage · Quaternary sedimentary deposits · Heritage education · Teachers' training programs

A. Domínguez-Almansa · T. Riveiro-Rodríguez (✉)
Faculty of Science Education, University of Santiago de Compostela, Santiago de Compostela, Spain
e-mail: tania.riveiro@usc.es

A. Domínguez-Almansa
e-mail: andres.dominguez@usc.es

R. Blanco-Chao · M. Costa-Casais
Faculty of History and Geography, University of Santiago de Compostela, Santiago de Compostela, Spain
e-mail: ramon.blanco@usc.es

M. Costa-Casais
e-mail: manuela.costa@usc.es

R. de Miguel González and P. Rivero Gracia (eds.), *Geospatial Technologies for Heritage Education*, Key Challenges in Geography,
https://doi.org/10.1007/978-3-031-96372-8_12

Introduction

This chapter advocates for the usefulness of introducing trainee teachers to teaching and learning processes that incorporate Geographic Information Systems (GIS). This set of tools is understood as an innovative element in the didactics of social sciences and is aligned with heritage education. To demonstrate the usefulness of GIS in this context, this chapter holistically addresses the study of space, territory, and a variety of places subject to processes of heritage appropriation.

This chapter proposes using GIS techniques and applications at various levels of depth. At the same time, it suggests applying these tools from an interdisciplinary and critical perspective, integrating physical geography and social knowledge, to address contemporary issues. This approach thus seeks to combine technical and interdisciplinary knowledge within a framework of socio-critical heritage education (Domínguez-Almansa et al. 2021).

The authors suggest that undergraduate or master's degree teacher training programs include the educational and social value of certain natural spaces, while being informed by the insights from heritage education and explored through the use of GIS. As a case study, we consider sedimentary deposits located along the coast of Galicia (north-western Spain) which, currently, function as active cliffs (Costa-Casais and Caetano Alves 2013). These sites hold significant scientific and educational value, as they provide valuable information on Earth's memory, recording past environmental and climatic changes (Blanco-Chao et al. 2019; Costas et al. 2009). However, despite the acknowledgement of their heritage value by the scientific community, these cliffs lack public awareness and recognition.

To identify, contextualise, catalogue, understand, and patrimonialise these visible records of past climate change—which allow critical reflexion on current climate change controversies—the use of GIS emerges as an important educational tool. Furthermore, the technical and scientific learning derived from this experience will enable its application to other natural or cultural spaces that, while unrecognised, may be of interest from an educational perspective.

Geographic Information Systems and Education in a Global Context

In education, GIS is a valuable tool for transforming the teaching and learning process of social sciences. Its application in the classroom enables the visualisation, manipulation, analysis, and understanding of real geospatial data, providing students and teachers with the opportunity to explore the world, examine recent social phenomena, and tackle future challenges in an innovative and interactive manner (Baker and White 2003; Muniz-Solari et al. 2015; Ross and Kerski 2005).

GIS has increasingly been integrated into educational curricula with positive results for students (Goldstein and Alibrandi 2013). Various proposals and strategies have been developed to incorporate it into classrooms, leveraging its capacity to foster not only a comprehensive geographical understanding—integrating spatial and social variables (Alibrandi and Sarnoff 2006)—but also critical thinking and the ability to address and solve social or environmental problems (Bearman et al. 2016; Demirci et al. 2013; Jo et al. 2016; Kerski 2003, 2008; Mayalagu et al. 2020; Patterson et al. 2003). Moreover, the use of GIS in the classroom aligns well with active and collaborative learning strategies and techniques (Bednarz 2004), which energise educational experiences that are set in local environments (Patterson et al. 2003).

Although the use of GIS is not widespread in classrooms internationally (Höhnle et al. 2016; Bernhäuserová et al. 2022), these positive results are also confirmed in isolated experiences in primary (Humble 2023; Jadallah et al. 2017) as well as secondary education, particularly in the case of Geography and History teachers (Chen and Wang 2015; DeMers 2016; De Miguel and De Lázaro Torres 2020; Rihelä and Mäki 2015; Sofias and Pierrakeas 2023). Similar results are also observed in university-level courses, especially those specialising in geography (Healy and Walshe 2019), or more general courses such as those focusing on teacher training (Harte 2017; Hong and Melville 2018; Mitchell et al. 2018; Walshe 2017). In this last area, numerous studies highlight the evolution of this tool's use, its educational potential, and its limitations when applied in the classroom. They also document relevant practical experiences and case studies (Garzón et al. 2019; Hong 2015; Mzuza and Van Der Westhuizen 2019; Walshe 2017).

Today, basic GIS options do not require complex technical training for non-expert use cases, such that teacher trainees can learn to handle them without any problems. However, more advanced options pose technical and methodological challenges when applied in the classroom (Collins and Mitchell 2018; Bernhäuserová et al. 2022). Therefore, adequate training, whether initial and/or ongoing, is essential during teacher training. Research shows that teachers who receive training in this area not only recognise the intrinsic value of GIS but also effectively incorporate it into their teaching (Walshe 2017). Furthermore, regardless of their subject area or educational level, they improve their ability and willingness to teach diverse topics (Lay et al. 2013; Mašterová 2023). Their technical proficiency, confidence, and willingness to use GIS in the classroom also increase particularly in practical and applied approaches (Hong 2014; Mašterová 2023).

GIS, Education, and Heritage

In the current context, the relationship between GIS, heritage, and education evolves with technological advancements and the more frequent implementation of innovative educational processes that consider this and other tools. There are studies on participatory mapping and documentation processes that facilitate the location and recording of historical monuments, archaeological sites, and cultural/natural landscapes. These processes provide a geospatial database that can be used not only for conservation purposes but also in the educational field if GIS is applied didactically (Coma Quintana and Rojo Ariza 2010; Crespo Castellanos 2019; Dasgupta et al. 2022; Lasorella et al. 2024; López 2022; Martínez-Hernández et al. 2024; Torres et al. 2017).

This interdisciplinary approach has proven effective in promoting the ability to analyse historical and geographical patterns across different levels of students, providing various learning pathways about the history and evolution of societies in both temporal and spatial dimensions (Coma Quintana and Rojo Ariza 2010). At other scales, working with various types of interactive systems, which are designed to facilitate understanding through their visual appeal, enables students to explore heritage spaces and sites in a different and effective way (Moreno et al. 2017). It also enables learning experiences focused on disseminating heritage beyond the school context (Buzo 2016; Velilla et al. 2021; Guallart et al. 2020). From this dissemination perspective, cooperative experiences between local communities and their schools are already being undertaken. Other initiatives, more focused on sensory experiences, propose using augmented and virtual reality as well as 3D models, introducing students to an alternative exploration of space and its landmark features (Carrera and Asensio 2016; Carrera et al. 2017).

Despite these experiences, there is still room for further research that links this triad's dimensions (GIS, heritage, and education) and focuses on natural heritage, as few studies address these dimensions, and even fewer do so through the use of GIS. This chapter, therefore, provides an educational proposal centred on natural spaces and intended for the socio-critical training of teachers.

Teacher Training, Natural Heritage, and GIS

Two complementary training options are proposed for future teachers. Initially, teachers must be trained in acquiring the necessary skills to use different software and manage data (Kerski 2003). Then, they should be provided with the tools and guidelines needed to develop teaching strategies that combine technical learning with research in their local environment. However, these two training options cannot be sustained without an intermediate layer of knowledge about the heritage, geographical, or social issues to be addressed in the classroom. Therefore, this chapter presents various training levels, using the Quaternary soil-sedimentary deposits of the Galician coast as reference, to integrate GIS into the study of social sciences and heritage education across different educational contexts. This learning experience is also intended to serve as a foundation for future proposals related to working with different natural landmarks or spaces.

Technical Skills Development: An Accessible Type of Training

The technical training and learning of future teachers regarding the use of GIS can and should develop along complementary levels.

(i) At an introductory level, the focus would be on using well-known viewers or 'web maps', such as Google Earth, or more specialised tools from the National Geographic Institute (MAPS Comparator, PNOA Orthophotos Comparator, or IBERPIX). Using these viewers, students learn to utilise GIS to understand the spatial and temporal dimensions of the subject matter presented in class. However, at this stage, teachers are not yet working directly with GIS. They would only have the ability to observe, compare, and interpret images included in GIS resources, without incorporating other data sources.

(ii) At an intermediate level, other capabilities within viewers such as IBERPIX would be utilised, particularly the ability to create simple mapping objects through vector drawing, and especially the option to import cartographic information layers, whether self-generated or obtained from external data sources. Thanks to Spain's Law on Spatial Information Infrastructure and Services (LISIGE; 14/2010) and the European INSPIRE Directive, Spain has a wealth of accessible cartographic information available via web map services (WMS) provided by various national and regional bodies (https://www.idee.es/), allowing many data sources to be easily incorporated without requiring knowledge of cartographic reference systems. Other open viewers, like SIGNA, function as full-fledged GIS programs, as they allow for more complex analyses using spatial and alphanumeric queries. However, both the interface and the mechanics of these queries require prior knowledge of GIS structure and functionality.

(iii) At an advanced level, the focus would shift to using software capable of digitising and exporting data as well as generating geospatial information (graphics and attributes), fully harnessing GIS' analytical capabilities. This level requires basic technical skills concerning data formats (vector formats like shapefiles or dgn, and raster formats), the use of WMS, and a minimum understanding of cartography (coordinate reference systems, scales, and spatial resolution).

From Technique to Knowledge

Technical training cannot be considered in isolation from knowledge of the subject being analysed. Future teachers must understand that technical skills alone are insufficient and must serve the broader purpose of knowledge acquisition. The focus here is on working within the framework of heritage education, utilising landmarks which, from geomorphological and geological perspectives, are considered natural heritage sites of great scientific value. In this case, Quaternary sedimentary deposits along the Galician coast offer, within their various layers, important information—whether visible or invisible—about the environmental changes of the last glacial period and the transition to the Holocene. Such formations provide an engaging educational context, as shown in previous studies (Domínguez-Almansa et al. 2021). On the one hand, they help reconstruct environmental conditions from past geological periods and connect them to human societies' activities and trajectories. On the other hand, they highlight their heritage value and the sense of identity and belong they inspire, especially given their relative abundance in the Galician landscape, and the great variety of facies they offer compared to other areas of the European Atlantic coast (Lieberman and Gordon 2021). This work thus involves both cataloguing and spatially classifying the deposits from a scientific and heritage perspective. A simple mobile device can geolocate and visualise photos, which can be later uploaded to free software like QGIS. In this way, questions related to their topography, aerial images, direct observations, or teacher explanations can be resolved (Figs. 1 and 2).

The activity's approach becomes more concrete during fieldwork and classroom analyses when using digital mapping. During the fieldwork, students are introduced to the object of study, covering basic concepts such as erosion, transport, sedimentation, or stratification. The significance of the deposits is explained, particularly as records of environmental changes in the planet's history, including climate conditions, and, in this case, variations in sea level.

Many of the factors that have contributed to the formation of these deposits can be analysed later using GIS viewers and digital cartography, especially by examining the

Fig. 1 Sedimentary deposit of Fonforrón (Porto do Son—Galicia). *Source* Own elaboration

Fig. 2 Students visiting the sedimentary deposit of Fonforrón (Porto do Son—Galicia). *Source* Own elaboration

topography of the area, the proximity of river systems, or the underlying geological substrate.

The classroom activity focuses on using cartographic information to answer questions posed during the fieldwork. Since a viewer with limited interactive analysis tools is used, the task primarily involves visually analysing different sources (topographic maps, aerial photographs, thematic maps, etc.). For example, questions might be asked such as, 'Where are the deposits located along the coast?' Through an initial, small-scale analysis, the deposits' positions within the estuary are identified, allowing for the analysis of their orientation and exposure to waves, the presence or absence of beaches, and considerations of potential threats such as the rise of sea levels or wave height, etc. On a larger scale, topography can be analysed using contour lines and topographic points, combined with shading models. The topographic analysis itself allows for the incorporation of other information, such as the articulation of the river network, geological features, changes in land use, and outlines of communication routes from aerial photos taken at different times, including older cartographic editions.

The wide variety of available sources and the study of the relationships between various natural factors and parameters (geology, topography, river networks, etc.) and human factors (land use, settlement patterns, etc.) can form a strong foundation for multidisciplinary work across various fields of knowledge (Fig. 3).

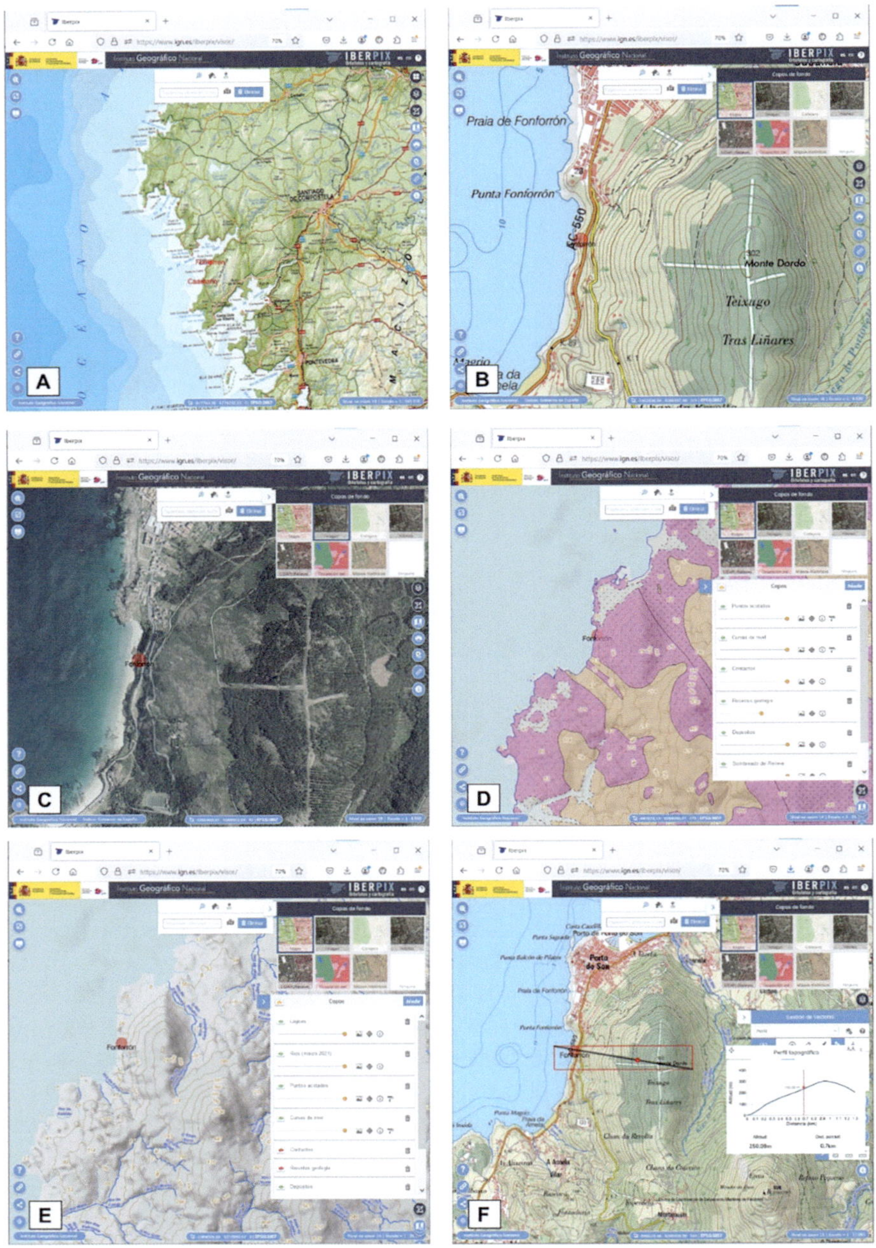

Fig. 3 Screenshots of the IBERPIX viewer. **A** Position of the studied deposits in the Muros and Noia rias; **B** Detail of the position and continental topography of the Fonforrón deposit; **C** Position of the deposit within the most up-to-date orthophotography; **D** Geological cartography from GEODE added via WMS; **E** Topographic representation using shading and contour lines from the IGN and the river network, added via the Xunta de Galicia's WMS; **F** Topographic profile obtained using the IBERPIX tools

Beyond the Specific Experience: Long-Term Lessons

Using this specific experience as a reference, trainee teachers would be given the opportunity to design and develop similar teaching activities that take space and its potential heritage elements as key components. The aim is that, independently, their lesson plan proposals for primary or secondary education would represent progress towards a more holistic and critical heritage education. Technical skills, combined with scientific and socio-critical knowledge, will enable the opening of processes for heritage appropriation, both in local settings and in other contexts. There should always be a special focus on fostering a teaching and learning process around natural landmarks with social, cultural, or historical significance. Forests, rock formations, peat bogs, lagoons, marshes, dunes, islands, and islets, among others, could be explored with the same parameters.

These experiences could be further reinforced by employing educational strategies that focus on the local environment, such as Place-Based Education or Enquiry-Based Learning. These, supported by collaborative work techniques and transversal civic objectives, tend to foster an emotional connection between students and the inhabited or studied space (Sobel 2005; Woodhouse and Knapp 2000). Ultimately, this bond aims to create a more sustainable and inclusive world where students engage with their communities and feel responsible for defending, preserving, and tending to their surroundings—ensuring that its natural value is not separated from its historical or social significance.

Conclusions

The proposed teaching model, rooted in a multidisciplinary approach encompassing social sciences and heritage education and focused on the appreciation and knowledge of natural landmarks, has the potential to reinforce the still limited and inconsistent understanding and use of geospatial technologies in classrooms (Baker et al. 2015).

This teaching model, through distinct phases of teacher training, (1) contributes to the educators' competency in GIS, understanding this technology as an innovative element; (2) connects the use of GIS to geospatial and historical thinking by working on territories and natural sites subject to heritage appropriation processes; (3) demonstrates this teaching model relationship with the curriculum and the learning possibilities for students capable of understanding physical geography and socio-critical knowledge holistically.

Given the wide range of benefits that GIS use brings to the teaching and learning process in the social sciences (De Miguel 2015), and in a context where new technological tools are emerging for experiential and localised territory analysis—such as cloud-based GIS or GIS viewers—incorporating the described approach into teacher training requires the continued demonstration of its utility in geography or history education (Buzo et al. 2022; Jo et al. 2016). Gradually, these experiences should be included in teacher training and normalised in future classrooms at various educational levels. For this, an update of training programmes as well as real-time support and follow-ups in classrooms are essential (Mitchell et al. 2018).

Studies show that both active teachers and their students have a positive attitude towards using GIS in the classroom, considering its presence insufficient yet expressing interest in greater proficiency and knowledge of available open-source software, available licenses, expanding curricula to include these resources, adapting school facilities to new realities, etc. (Šiljeg et al. 2022). This is coupled with the growing interest physical geography has received in recent years, as its potential has been recognised in a constructivist and meaningful teaching and learning process that fosters students' understanding of how Earth works and its implications for history and everyday life (Day 2012).

In this regard, future trainee teachers have shown interest in holistic learning derived from the discovery and understanding of soil-sedimentary deposits (Domínguez-Almansa et al. 2021). The promotion of GIS tools could be well-received as an extension of this educational experience, while considering it both a technical and educational challenge that might require greater collaboration among university faculty from various disciplines.

Acknowledgements This research was partly financed through project ED431C 2022/09 (Xunta de Galicia-Spain).

The four authors are members of the research group: Environmental Studies applied to Natural and Cultural Heritage of the University of Santiago de Compostela (GEMAP—GI 1243).

Ramón Blanco-Chao and Manuela Costa-Casais are also members of the Interuniversity Research Centre for Atlantic Cultural Landscapes (CISPAC).

References

Alibrandi M, Sarnoff HM (2006) Using GIS to answer the whys of where in social studies. Soc Educ 70(3):138–143

Baker TR, Battersby S, Bednarz SW, Bodzin AM, Kolvoord B, Moore S, Sinton D, Uttal D (2015) A research agenda for geospatial technologies and learning. J Geogr 114(3):118–130. https://doi.org/10.1080/00221341.2014.950684

Baker TR, White SH (2003) The effects of GIS on students' attitudes, self-efficacy, and achievement in middle school science classrooms. J Geogr 102(6):243–254. https://doi.org/10.1080/00221340308978556

Bearman N, Jones N, Andre I, Cachinho HA, DeMers M (2016) The future role of GIS education in creating critical spatial thinkers. J Geogr High Educ 40(3):394–408. https://doi.org/10.1080/03098265.2016.1144729

Bednarz S (2004) Geographic information systems: a tool to support geography and environmental education? GeoJournal 60(2):191–199 https://doi.org/10.1023/B:GEJO.0000033574.44345.c9

Bernhäuserová V, Krajňáková L, Hátlová K, Hanus M (2022) The limits of GIS implementation in education: a systematic review. ISPRS Int J Geo Inf 11(12):592. https://doi.org/10.3390/ijgi11120592

Blanco-Chao R, Costa-Casais M, Cajade-Pascual D, Gómez Rey G (2019) Coastal retreat and sedimentation during the last 3000 years. Atlantic Coast of NW Spain. J Mar Sci Eng 7(10):331. https://doi.org/10.3390/jmse7100331

Buzo Sánchez I (2016) Aplicación de la Metodología del Aprendizaje Geográfico por descubrimiento basado en SIG en proyectos didácticos para 2º de Bachillerato. In Sebastià Alcaraz and Emilia Tonda Monllor, La investigación e innovación en la enseñanza de la Geografía. UNE, pp 477–489

Buzo I, Mínguez C, De Lázaro ML (2022) Expert perspectives on GIS use in Spanish geographic education. Int J Digit Earth 15(1):1204–1218. https://doi.org/10.1080/17538947.2022.2096131

Carrera CC, Asensio LAB (2016) Augmented reality as a digital teaching environment to develop spatial thinking. Cartogr Geogr Inf Sci 44(3):259–270. https://doi.org/10.1080/15230406.2016.1145556

Carrera CC, Avarvarei BV, Chelariu EL, Draghia L, Avarvarei SC (2017) Map-reading skill development with 3D technologies. J Geogr 116(5):197–2005. https://doi.org/10.1080/00221341.2016.1248857

Chen CH, Wang YH (2015) Geospatial education in high schools: curriculums, methodologies, and practices. In: Solari IOM, Demirci A, van der Schee J (eds) Geospatial technologies and geography education in a changing world: geospatial practices and lessons learned. Springer, pp 67–76

Collins L, Mitchell JT (2018) Teacher training in GIS: what is needed for long-term success? Int Res Geogr Environ Educ 28(2):118–135. https://doi.org/10.1080/10382046.2018.1497119

Coma Quintana L, Rojo Ariza MC (2010) Arqueología y museografía didáctica en los aeródromos de Guerra (1936–1939). Ebre 38(4):165–177

Costa-Casais M, Caetano Alves IM (2013) Geological heritage at risk in NW Spain. Cuaternary deposits and landforms of Southern Coast (Baiona-A Garda). Heritage 5:227–248. https://doi.org/10.1007/s12371-013-0083-7

Costas S, Sobrino CM, Alejo I, Pérez-Arlucea M (2009) Holocene evolution of a rock-bounded barrier-lagoon system, Cies Islands, northwest Iberia. Earth Surf Proc Land 34(11):1575–1586. https://doi.org/10.1002/esp.1849

Crespo Castellanos JM (2019) El visualizador Iberpix 4 del Instituto Geográfico Nacional: un recurso didáctico para la interpretación de los componentes físicos del paisaje. Enseñanza de las Ciencias de la Tierra 27(2):182–191. https://raco.cat/index.php/ECT/article/view/367134

Dasgupta R, Hashimoto S, Basu M, Okuro T, Johnson BA, Kumar P, Dhyani S (2022) Spatial characterization of non-material values across multiple coastal production landscapes in the Indian Sundarban delta. Sustain Sci 17(3):725–738. https://doi.org/10.1007/s11625-020-00899-3

Day T (2012) Undergraduate teaching and learning in physical geography. Prog Phys Geogr Earth Environ 36(3):305–332. https://doi.org/10.1177/0309133312442521

DeMers MN (2016) Geospatial technology in geography education. Geogr Teach 13(1):23–25. https://doi.org/10.1080/19338341.2016.1151722

De Miguel GR (2015) Del pensamiento espacial al conocimiento geográfico a través del aprendizaje activo con tecnologías de la información geográfica. Revista Geografia do Colegio Pedro II 2(4):13

De Miguel GR, De Lázaro Torres ML (2020) WebGIS implementation and effectiveness in secondary education using the digital atlas for schools. J Geogr 119(2):74–85. https://doi.org/10.1080/00221341.2020.1726991

Demirci A, Karaburun A, Ünlü M (2013) Implementation and effectiveness of GIS-based projects in secondary schools. J Geogr 112(5):214–228. https://doi.org/10.1080/00221341.2013.770545

Domínguez-Almansa A, Costa-Casais M, Riveiro-Rodríguez T (2021) Cliffs with memory: historical and social teacher training via climatic and environmental transformations. In: Carrasco CJG, Martínez PM, Facal RL (eds) Handbook of research on teacher education in history and geography. Peter Lang, pp 331–354. https://doi.org/10.3726/b18091

Garzón A, Rodríguez JM, Cara B, López-Carrique E (2019) The use of heographical information systems (GIS) to teach sciences in secondary education. In: Chova LG, Martínez AL, Torres IC (eds) 13th international technology, education and development conference (INTED). INTED proceedings, pp 241–248

Goldstein D, Alibrandi M (2013) Integrating GIS in the middle school curriculum: impacts on diverse students' standardized test scores. J Geogr 112(2):68–74. https://doi.org/10.1080/00221341.2012.692703

Guallart C, Vevilla J, Cuartero N, Ferraz MF, Laguna M, Ollero A, Rodrigo B (2020) Propuesta de recursos didácticos en línea para trabajar patrimonio, despoblación y territorio en Educación Secundaria Obligatoria. Cuadernos de Geografía 104:153–176. https://doi.org/10.7203/CGUV.104.16783

Harte W (2017) Preparing preservice teachers to incorporate geospatial technologies in geography teaching. J Geogr 116(5):226–236. https://doi.org/10.1080/00221341.2017.1310274

Healy G, Walshe N (2019) Real-world geographers and geography students using GIS: relevance, everyday applications and the development of geographical knowledge. Int Res Geogr Environ Educ 29(2):178–196. https://doi.org/10.1080/10382046.2019.1661125

Höhnle S, Fögele J, Mehren R, Schubert JC (2016) GIS teacher training: empirically-based indicators of effectiveness. J Geogr 115(1):12–23. https://doi.org/10.1080/00221341.2015.1016546

Hong JE (2015) Teaching GIS and other geospatial technologies to in-service teachers. In Solari OM, Demirci A, van der Schee J (eds) Geospatial technologies and geography education in a changing world: geospatial practices and lessons learned. Springer, pp 117–126. https://doi.org/10.1007/978-4-431-55519-3_10

Hong JE (2014) Promoting teacher adoption of GIS using teacher-centered and teacher-friendly design. J Geogr 113(4):139–151. https://doi.org/10.1080/00221341.2013.872171

Hong JE, Melville A (2018) Training social studies teachers to develop inquiry-based GIS lessons. J Geogr 117(6):229–244. https://doi.org/10.1080/00221341.2017.1371205

Humble N (2023) Future paths for GIS in K-12 education: a review of possibilities and constraints. European J Geogr 14(3):79–88. https://doi.org/10.48088/ejg.n.hum.14.3.079.088

Jadallah M, Hund AM, Thayn J, Studebaker JG, Roman ZJ, Kirby E (2017) Integrating geospatial technologies in fifth-grade curriculum: impact on spatial ability and map-analysis skills. J Geogr 116(4):139–151. https://doi.org/10.1080/00221341.2017.1285339

Jo I, Hong JE, Verma K (2016) Facilitating spatial thinking in world geography using web-based GIS. J Geogr High Educ 40(3):442–459. https://doi.org/10.1080/03098265.2016.1150439

Kerski JJ (2008) Developing spatial thinking in education and society. https://www.researchgate.net/publication/267365595_Developing_Spatial_Thinking_Skills_in_Education_and_Society

Kerski JJ (2003) The implementation and effectiveness of geographic information systems technology and methods in secondary education. J Geogr 102(3):128–137. https://doi.org/10.1080/00221340308978534

Lasorella M, de-Dato P, Cantatore E (2024) Web-based platform to collect, share and manage technical data of historical systemic architectures: the telegraphic towers along the Madrid-Valencia path. Virtual Archaeol Rev 15(30):123–140. https://doi.org/10.4995/var.2024.20341

Lay JG, Chi YL, Hsieh YS, Chen YW (2013) What influences geography teachers' usage of geographic information systems? A structural equation analysis. Comput Educ 62:191–195. https://doi.org/10.1016/j.compedu.2012.10.014

Lieberman B, Gordon E (2021) El cambio climático en la historia de la humanidad. Desde la prehistoria al presente. Ed Almuzara

López Sández M (2022) La música popular y la educación literaria en contextos multiplingües. Una propuesta interartística y multicultural con incorporación de Sistemas de Información Geográfica. Cauce: Revista Internacional de Filología, Comunicación y sus Didácticas 45(1). https://doi.org/10.12795/CAUCE.2022.i45.04

Martínez-Hernández C, Stoffelen A, Piskorski R (2024) Obtaining geographical competences through online cartography of familiar and unfamiliar urban heritage: lessons from student workshops. Geogr High Educ 48(1):74–93. https://doi.org/10.1080/03098265.2022.2155935

Mašterová V (2023) Learning and teaching through inquiry with geospatial technologies: a systematic review. Eur J Geogr 14(3):42–54. https://doi.org/10.48088/ejg.v.mas.14.3.042.054

Mayalagu G, Jaafar M, Lam Kuok C, Mahmud MI (2020) The effectiveness of geographic information system (GIS) on spatial thinking skills (STS) among geography students. Int J Adv Sci Technol 29(9):1000–1009

Mitchell JT, Roy G, Fritch S, Wood B (2018) GIS professional development for teachers: lessons learned from high-needs schools. Cartogr Geogr Inf Sci 45(4):292–304. https://doi.org/10.1080/15230406.2017.1421482

Moreno Nevado M, Becerra Ramírez R, González Luna ME (2017) Experiencias de innovación docente en enseñanza superior de Castilla-La Mancha, 2017. Procedings: II Jornada de Innovación Docente Universidad de Castilla-La Mancha. Albacete, 28 junio 2017 / coord. Mabel López Solera, Ana María Sanz Redondo, and Caridad Pérez de los Reyes, pp 25–26

Muniz-Solari O, Demirci A, Van der Schee J (2015) Geospatial technologies and geography education in a changing world. Springer. https://doi.org/10.1007/978-4-431-55519-3

Mzuza MK, Van Der Westhuizen CP (2019) Review on the state of GIS application in secondary schools in the southern African region. S Afr Geogr J 101(2):175–191. https://doi.org/10.1080/03736245.2019.1579110

Patterson MW, Reeve K, Page D (2003) Using GIS to enhance students' understanding of complex environmental and social issues. J Geogr 102(6):275–281. https://doi.org/10.1080/00221340308978559

Rihelä J, Mäki S (2015) Designing and implementing an online GIS tool for schools: the finnish case of the PaikkaOppi Project. J Geogr 114(1):15–25. https://doi.org/10.1080/00221341.2014.897362

Ross S, Kerski JJ (2005) The essentials of the environment: understanding the impact of human activities on the natural world. Routledge

Šiljeg S, Milanovic A, Maric I (2022) Attitudes of teachers and students towards the possibilities of GIS implementation in secondary schools in Croatia. Educ Sci 12(12):846. https://doi.org/10.3390/educsci12120846

Sobel D (2005) Connecting classrooms and communities, 2nd ed. Orion Society

Sofias TA, Pierrakeas CJ (2023) Student engagement and educational benefits of Web GIS-based projects. Int J Web Based Learn Teach Technol 18(1):16. https://doi.org/10.4018/IJWLTT.317089

Torres MLD, González RD, Yago FJM (2017) WebGIS and geospatial technologies for landscape education on personalized learning contexts. ISPRS Int J Geo Inf 6(11):350. https://doi.org/10.3390/ijgi6110350

Velilla Gil J, Guallart Moreno C, Marin-Yaseli ML (2021) Educational materials to work on sustainable development concepts in secondary education classrooms. GEOFOCUS-Revista Internacional de Ciencia y Tecnologia de la Información Geográfica 27:55–93. https://doi.org/10.21138/GF.676

Walshe N (2017) Developing trainee teacher practice with geographical information systems (GIS). J Geogr High Educ 41(4):608–628. https://doi.org/10.1080/03098265.2017.1331209

Woodhouse JL, Knapp CE (2000) Place-based curriculum and instruction: outdoor and environmental education approaches. ERIC Digest. https://eric.ed.gov/?id=ED448012

Geotextual Analysis of Conflict Heritage Narratives in Museums: The Case of the Battle of the Ebro (1938)

Xavier Rubio-Campillo, Tania González-Cantera, and Giulio Roberto Bartuli

Abstract Historical conflicts have radically transformed our landscapes through the construction of fortifications, the destructive impact of weapons, and the post-war memorialization processes. Therefore, battlefields function as both archaeological sites and heritage locations since they serve as valuable educational resources about the impact of violence on society. Battlefield museums and conflict heritage institutions play a crucial role in presenting the context and consequences of warfare to the public. Analysing these representations can reveal insights into the perception of conflict landscapes and their significance in the official narratives of traumatic events. This work presents an innovative methodological approach to the geography of historical conflict grounded on Digital Humanities. This approach was applied to analyse the geography of conflict of one of the most decisive and largest battles ever fought in Spain: the Battle of the Ebro (1938). The region where this costly battle took place is known as Terra Alta, and as its name suggests ('Highlands' in Catalan), it is crowded with mountain ranges that facilitated the Republican defence against the material superiority of the Francoist army. This region currently hosts several museums themed on the battle; data collection was performed to create a corpus of the texts displayed in the permanent exhibition of six of these museums. This corpus was further digitized and integrated into a relational database. The corpus was then annotated by identifying and georeferencing the place names found in the texts to be used within a Geographic Information System (GIS). The dataset was also explored by using a combination of spatial analysis and data visualization methods to define

X. Rubio-Campillo (✉)
Departament de Didàctiques Aplicades, Universitat de Barcelona, Barcelona, Spain
e-mail: xrubio@ub.edu

Institut de Recerca en Arqueologia (IAUB), Universitat de Barcelona, Barcelona, Spain

T. González-Cantera
Departament d'Història i d'Història de l'Art, Universitat de Girona, Girona, Spain
e-mail: taniagonzalezcantera@gmail.com

G. R. Bartuli
Universitat de Barcelona, Barcelona, Spain
e-mail: giuliorobertobartuli@gmail.com

R. de Miguel González and P. Rivero Gracia (eds.), *Geospatial Technologies for Heritage Education*, Key Challenges in Geography,
https://doi.org/10.1007/978-3-031-96372-8_13

the places that are more visible in the narratives of the museum as well as to identify missing or forgotten places. Results generated a rich and complex portrayal of how geography shaped the battle and what landscape elements have become essential in the narratives taught to the visitors. The interpretation of these findings reveals the role played by geography in heritage presentation and peace education, and how landscape is used to build official narratives aimed at educating about such historical traumatic events as the Spanish Civil War.

Keywords Spanish Civil War · Digital Humanities · Geotextual analysis · Geographic Information Systems · Conflict heritage

Introduction

A historical battlefield is one of the most complex types of landscapes, understood as "an area, as perceived by people, whose character is the result of the action and interaction of natural and/or human factors" (Europe 2000). This definition implies that a battlefield is not just a geographical site of extreme violence, but also reflects the diverse relationships people have with that space over time before and after combat. In this complex dynamic, the battle shapes the landscape by altering it significantly, marked by structures for attack and defence (such as trenches and machine gun nests), remnants of conflict (e.g., scars from aerial bombardments), and landmarks for remembrance (memorials, plaques, cemeteries). Similarly, the geographical context also shapes the battle, by determining both military strategies and the experiences of those involved; living war in front-line trenches differs greatly from enduring it in the rearguard. Conflict is certainly one of the most intense and traumatic events for any society and particularly for the people who experienced it. For this reason, battlefield landscapes or battlescapes (Kolberg 2023; Rotherham 2024) have become symbols of self-identity for both participants and future generations.

The Battle of the Ebro (July–November 1938) is one of the most important battlefields in European history due to the scale of combat that took place and the political significance of its outcome within the Spanish Civil War (1936–1939). The defeat of the Republican army after 100 days of combat and tens of thousands of casualties was one of the main reasons behind the victory of Francisco Franco which led to four decades of dictatorship in Spain (Martínez Bande 1988; Martínez Reverte 2003). The best combat units of both armies collided in a bloody engagement that spanned several months, leaving a lasting mark on the landscape. Two factors significantly influenced the battle's course: the massive use of aviation, particularly from the Francoist air force and its Italian and German allies, and the geography of the region, known as Terra Alta (Highlands) which was dominated by mountain ranges such as Pàndols and Cavalls as well as peaks where most combats took place (Rubio-Campillo and Hernàndez 2015).

The memory of the battle has been preserved and visibilized by a corpus of published works collecting the experience of the participants (Cardona and Losada

2004; Estrada Vidal 1973; Henríquez Caubín 2009), but this memory is also written in the landscape. Today, the Ebro's battlescape is a palimpsest of scars and wounds that serves as a constant reminder of the brutal battle that took place 85 years ago. These scars have not only left an imprint on the landscape, but also have shaped the identity of the region. This is reflected in the various organizations and museums in the Ebro area dedicated to studying and preserving the memory of the battle; in fact, the Ebro is by far the Spanish Civil War battlefield with a most intense memorialization process, including the creation of several museums and visitor centres.

Following some pioneering local initiatives to commemorate the Battle of the Ebro, the Catalan government established the Consorci d'Espais de la Batalla de l'Ebre in the 2000s (COMEBE—Consortium of Places of the Battle of the Ebro). COMEBE is a network of museums created to institutionalize the remembrance of the battle (González Vázquez 2016). This consortium manages 5 museums, 22 historical spaces, and 1 memorial within the Terra Alta region. The huge size of the Ebro battlescape means that its memorialization beyond museum exhibitions remains a challenge for the region. The preservation of the landscape as historical heritage is often seen as a secondary goal compared to initiatives with direct economic gain such as the construction of large-scale wind parks in the same mountain ranges where the battle took place (Busquets and Ramos 2019), as well as political and ideological debates on the creation of a unified memory of the battle (Sospedra-Roca et al. 2023, p. 465). These tensions have sparked public controversies, often linked to the struggle to recover the memory of the war (González Ruibal 2020; Rubio-Campillo and Hernàndez 2015). As a result, a diverse set of initiatives emerged outside the COMEBE umbrella as some local associations have created museums other perspectives on the Battle of the Ebro's heritage, including the Memorial Museum of the Battle of the Ebro (Museu Memorial de la Batalla de l'Ebre) in Gandesa (Catalonia) and the Museum of the Battle of the Ebro (Museo de la Batalla del Ebro) in Faió ('Fayón' in Spanish; Aragon).

This research aims to identify the role that the Ebro battlescape plays in the heritage presentation of this crucial battle in European history. A Digital Humanities novel method combining geotextual analysis, spatial analysis, and data visualization has been developed to explore the geographical dimension of these narratives and to what extent the permanent exhibitions capture the complexity and importance of the Terra Alta geography within the historical event.

This chapter is structured as follows: next section defines the steps of this innovative methodology with special attention to the geospatial and spatial analyses. The results of location frequency, place co-occurrences and spatial analysis are then presented, followed by an in-depth discussion of the relationship between the Ebro battlescapes and the narratives of its museums.

Materials and Methods

The method developed to explore the role of the Ebro landscape in the narratives of the battle is a novel computational approach grounded on recent trends using geotextual analysis (Alex et al. 2019; Gregory et al. 2015; Murrieta-Flores and Martins 2019). The section defines the sequence of steps that can be summarized as follows: (a) creation of a corpus of heritage-related texts, (b) georeferentiation of corpus based on place names, (c) identification of place name frequencies and co-locations, and (d) assessment of topographical traits of the battlescape including visibility and accessibility indexes.

Place Name Annotation

The starting point of the analysis is the creation of a corpus that included the textual panels of the Ebro museums. Six museums were visited during early 2024 from which all textual panels were photographed: CI 115 dies, CI Veus del Front, CI Internacionals a l'Ebre, and CI Hospitals de Sang from the COMEBE network (we left out CI Soldats a les Trinxeres because it is currently closed) and two museums with associative-origin: Museu de la Batalla de l'Ebre (CEBE) in Gandesa, and Museu de la Batalla del Ebro in Faió). The analysis focused on permanent exhibitions given that this is the most stable source of information that the visitor receives. In the case of the Faió museum data collection also included the text of the audio guides which were kindly provided by the staff. The reason behind this decision was that the permanent exhibition is being remodelled and the new texts would be based on the script of the guides.

The second step was to transform the content into text format. Optical Character Recognition (OCR) software was used on the hundreds of photographs taken in the different museums. OCR is a popular technology that identifies text in rasterized images and converts it to a machine-readable format which can be used in further analysis (for an overview on the digitization process, see Jentsch and Porada 2021). Google Keep was the OCR tool chosen for this work; although its primary function is to take notes, make lists, and create reminders, this application also includes OCR software that allows transcribing the text present in the uploaded photos and copying it into different formats (e.g., text file, word processing software).

OCR software typically has a high accuracy rate, but some errors were inevitable given the complexity of panel texts compared to other sources of text (e.g., unusual fonts, low-quality lighting conditions, and visual elements superimposed on the text). Consequently, the team carried out a manual revision on the entire dataset and, thus, common errors—such as incorrect character recognition, omission of words or phrases, and incorrect paragraph segmentation—were corrected. The structure was also normalized to facilitate analysis, thus removing line breaks or complex spacing that may have been introduced by the OCR process. The final dataset included text of 38 panels and an overall sample of over 43.000 words. This corpus was in Catalan language, as it was the original language of the texts (some of them also included Spanish and English translations).

The last phase of the data creation process was the identification of place names within the museum corpus. This identification was done using text annotation, which is a technique used to link semantic tags to text fragments to define their content (Rubio-Campillo 2023). Text annotation can be manual or automated, and it is a key component of present-day Digital Humanities practices which has generated important methodological efforts such as the Text Encoding Initiative consortium (Burnard 2013). The text annotation tool used in this work is the Historical Textual Navigator (HisTeNa) developed by the main author of this work to facilitate annotation of historical sources (Rubio-Campillo 2023). This software couples an intuitive graphical interface for manual annotation with a SQLite database specifically designed for small-sized historical research projects such as this one.

HisTeNa allows users to code three types of tags: (a) persons, (b) custom keywords, and (c) place names. The encoding of place names allows the user to select locations from a list of homonymous places from the open GeoNames database. If the place name is not found then the user can manually create a new entry including name, description, latitude, and longitude coordinates. This detail is important while working on historical documents because key place names on these documents may not be currently in use, or may have changed their spatial position. As an example, one of the key towns of the battle is Corbera d'Ebre; after the war, the place was abandoned due to the destruction caused by the combats, and a new town was built 500 m from the old town. A new place name was created (called Poble Vell, Corbera d'Ebre old town) and positioned in the abandoned town instead of the one that can be found in GeoNames.

The geotextual annotation of the corpus identified 645 place names and required the definition of 64 new place names including military positions and peak identifications used by both armies (i.e., 'cotes' in Catalan) that had to be georeferenced using historical documents and cartography.

Geotextual Analysis

Computational text-based analysis was applied to the corpus to assess the presence of geography within the narratives of the battle by focusing on the frequency and relations of place names. This analysis was developed using a diversity of data visualization methods. Data visualization is a statistical technique that allows researchers to reveal complex patterns on multivariate datasets (Cairo 2012). This visual approach is an effective strategy for Digital Humanities works because it provides a good balance between quantitative and qualitative methods. On the one hand, modern graphs can provide deep insights combining multiple variables within the same figure; on the other hand, data visualization facilitates interpretation considering any bias identified by the experts (Rubio-Campillo 2023). Two main strategies have been used in this work (Cao and Cui 2016):

- Barplots have been used to compare the frequency of place names both over the entire dataset and split by museum.
- Relations between place names have been identified by transforming a matrix of term co-locations into a network of interaction.

Spatial Analysis

The study of battlescapes needs to compare the relevance of its landscape during the historical events with its presence in the narratives generated after them grounded on present-day existing heritage. To this end, spatial analysis was performed to assess two essential traits for any conflict landscape: visibility and mobility.

The visual control of the battlefield is vital for any army, as it allows its commanders to assess the strength of the positions of the enemy, to follow the development of combat, and to guide the accuracy of indirect fire (i.e., fire weapons that do not have a direct line of sight to enemy positions) from artillery. Visibility was even more relevant in the battle of the Ebro, as the landscape was characterized by several mountain chains and rough terrain; in fact, the Republican Commander, Vicente Rojo, deliberately chose this battlefield because rough terrain would facilitate defence, thus minimizing the impact of the material superiority of the Francoist army, especially in terms of artillery and bombers (Alonso Baquer 1988).

A visibility index for the entire battlefield was computed as a percentage of total viewshed. The algorithm computed for each pixel in a 5-m resolution raster the percentage of neighbouring cells within a radius that could be observed from it. The radius for this case was defined as 5 km, since it is a reasonable distance where enemy positions can be tracked in good weather. The output of this algorithm is a raster where each cell has a value between 0 (no neighbours can be seen) and 1 (all neighbours are observed).

Mobility analysis was performed with Circuitscape (Anantharaman et al. 2020; McRae et al. 2008). This tool applies circuit theory to spatial information to identify regions of high connectivity while moving between different areas as well as bottlenecks that may hinder mobility. Circuitscape was developed for research on ecology, but it has been recently applied to historical landscapes due to its flexibility and its capability to generate general overviews on the mobility of an entire region (Howey 2011; McLean and Rubio-Campillo 2022; Thayn et al. 2016). The approach taken here was to define an origin region including the Ebro riverside from Faió to Benifallet and a destination covering the road exits at the western section of the map (i.e., the Republican objectives). The algorithm identified the most probable regions to move from origin to destination taking into account an isotropic cost function based on terrain slope (see Rubio-Campillo et al. 2022 for the computational details).

The final step was to combine both indexes with data visualization methods; sampling was performed on both raster maps at the locations of the identified place names and then displayed using a scatter plot.

Results

Location Frequency

The first outcome of geotextual analysis is focused on the most popular locations found in the museums of the battle. Figure 1 displays the results of the frequency analysis for the entire corpus as well as each museum as a bar chart depicting the top 10 most mentioned place names.

This graph is complemented by the geographical distribution of place names. Figure 2 displays a general map of the Iberian Peninsula including all place names mentioned in the corpus which allows us to assess the historical and geographical context provided to the visitor beyond the Ebro. This overview is complemented by Fig. 3 which provides a zoomed-in map of the battlefield which facilitates the understanding of the battlescape.

The combination of the three figures shows that there are some mentions to earlier offensives of the Spanish Civil War to contextualize the Ebro battle (e.g., Brunete, Teruel, Valencia). However, most place names fall within the Terra Alta region, where the battle took place. The most mentioned battlefield locations in the museum narratives are the towns where the Republican offensive was stopped by the Francoist army (e.g., Faió, Gandesa, Vilalba dels Arcs), followed by additional towns and landmarks that saw major combat (e.g., Corbera d'Ebre, Venta de Camposines).

Place Co-ocurrences

The relation between place names was explored by assessing co-occurrences, understood as the frequency at which any given pair of place names of the dataset are mentioned within the same document (in our case text panel). This information can be understood as a network of links where each node is a place name and the co-ocurrence between two of them is seen as an edge. Figure 4 shows this network of interactions for the most frequent co-ocurrences of locations (e.g., place names mentioned together at least four times).

The central location of the network is dominated by the towns of Gandesa and Corbera d'Ebre (where CEBE and *CI* 115 dies are currently located) and Barcelona as the main city in the Republican rearguard in Catalonia. In contrast, Faió (i.e. the most repeated place name in the dataset) is far from the centre of the network and mainly connected to place names of the Faió-Mequinensa pocket.

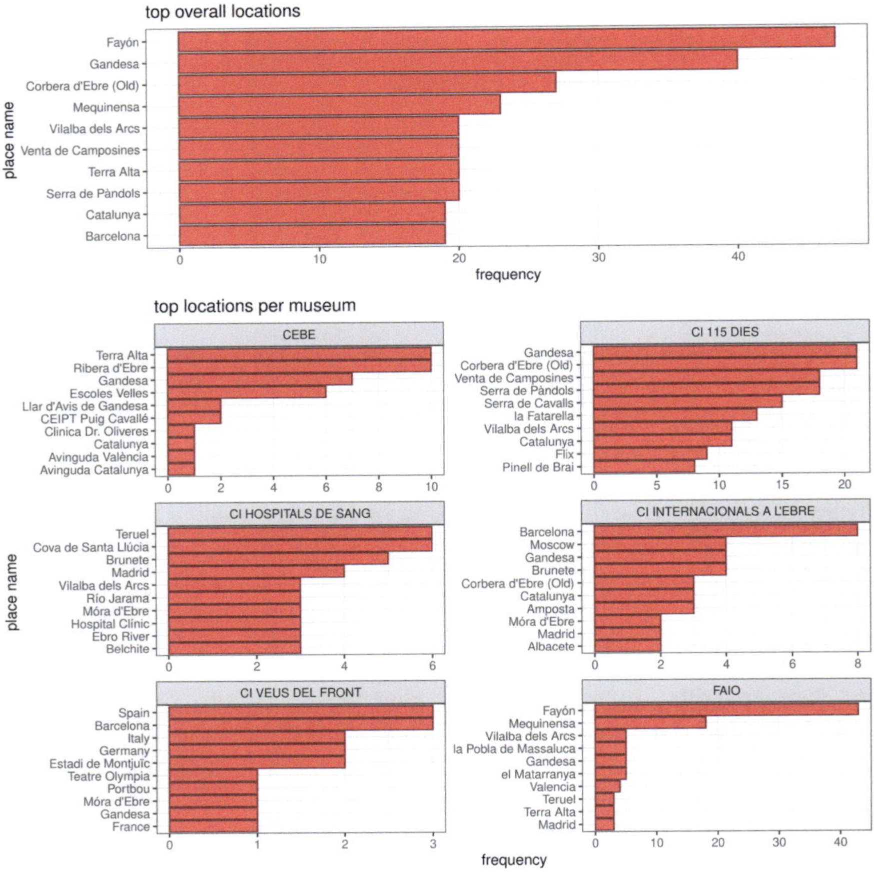

Fig. 1 Top 10 most frequent locations found in the museums explaining the battle of the Ebro. Top image shows overall frequencies while bottom is split by museum

Spatial Analysis

Figure 5 presents the results of the visibility analysis: Terra Alta's roughness generates highly fragmented viewsheds because lines of sight are broken by the large number of peaks, hills, and mountains. Therefore, the higher visibility regions in the battlescape do not go higher than 0.2 (range 0–1) except for the areas adjacent to the Ebro river, where some plains exist (e.g., Móra d'Ebre) and the peaks of the mountain ranges that dominate the Venta de Camposines-Gandesa road.

Figure 6 shows the output of connectivity analysis. Natural routes follow the valleys connecting the Ebro River with the Western part of the region as well as some secondary routes linking the Southern part of the battlefield with these primary routes. It is worth noting that present-day roads closely follow the regions of high connectivity.

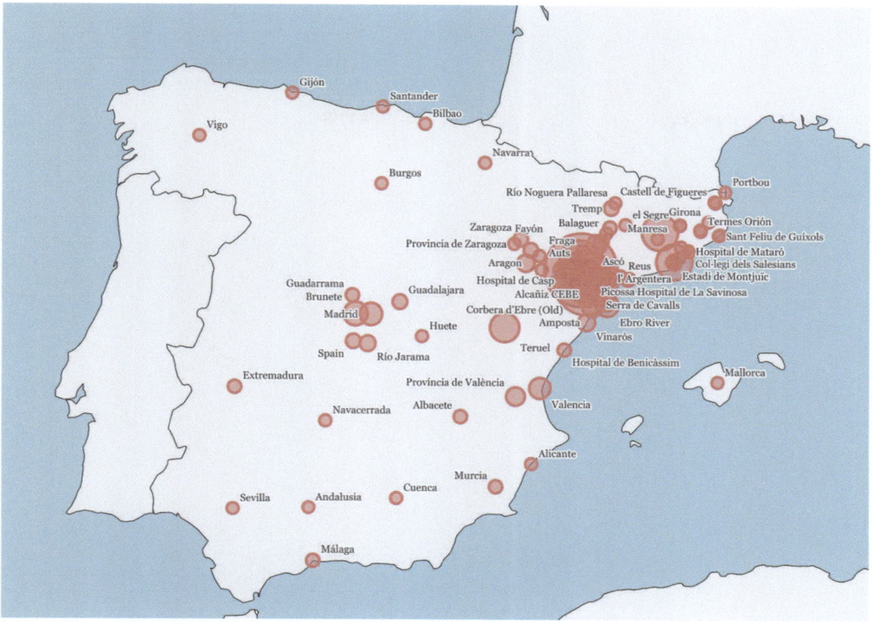

Fig. 2 Map of general georeferenced locations found in the dataset of museums. The size of each place is proportional to its frequency

Finally, in Fig. 7 we compare these two spatial indexes using a scatterplot to identify clusters of place names with specific correlations between visibility and connectivity.

Discussion

Place name frequencies show that there exists a split between Faió and the rest of the museums: the most frequent place names in Faió are Faió itself and Mequinensa, while these two locations are seldom referenced in the rest of the dataset. This split is the consequence of the physical division between the Faió pocket which established between these two towns and the rest of the battlefield. The Faió pocket was eliminated during the first Francoist August counteroffensive, and for this reason, it played a minor role in the final outcome of the battle. Nonetheless, the Faió museum is focused on explaining this secondary action given its spatial proximity.

Another geographical insight linked to the narratives is the similarity of place names in *CEBE* and *CI 115 dies*, the largest museums and the ones trying to capture the entire battle. In contrast, the rest of the COMEBE network explores different themes and, for this reason, each of them use a distinct catalogue of place names. For example, *CI Hospitals de Sang* in Batea explains the vital role played by hospitals

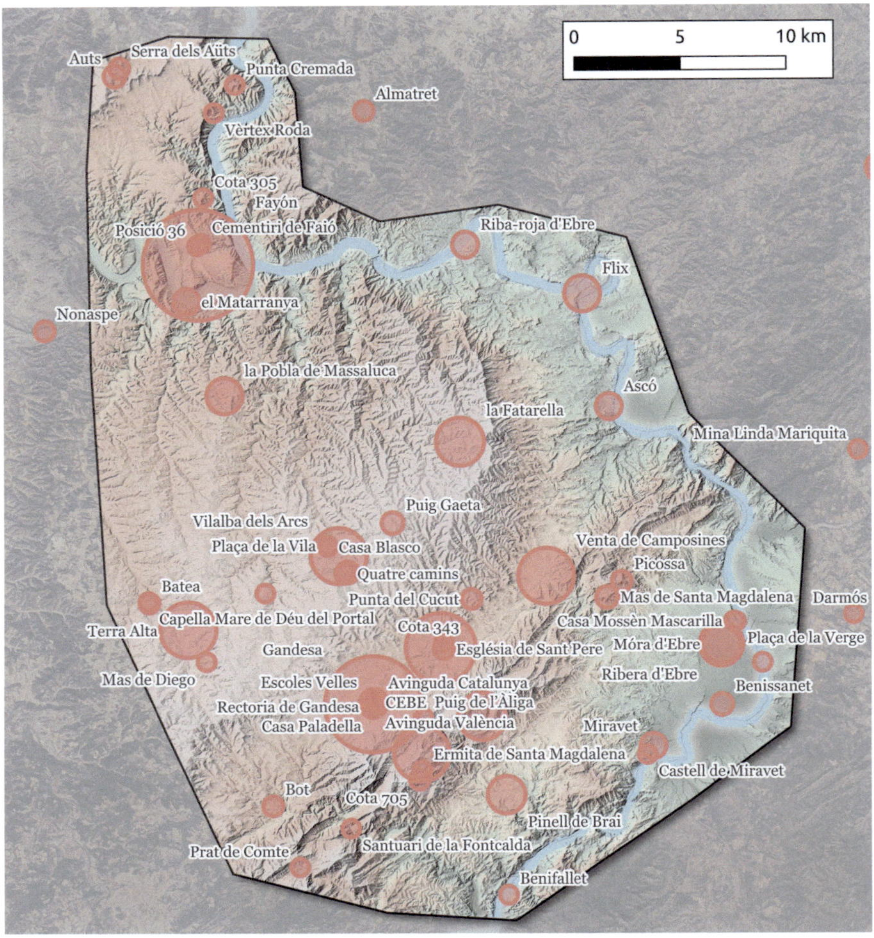

Fig. 3 Map of georeferenced locations within the region of the battlefield. The size of each place is proportional to its frequency

to treat wounded soldiers coming from the front, and it mentions locations where casualties were sent (e.g. Cova de Santa Llúcia, Hospital Clínic). Similarly, *CI Veus del front* is focused on the press and international public opinion, thus mentioning rearguard cities and countries instead of battlescape locations.

A key outcome of the analysis is that the museum narratives are much more focused on populated locations than on the geographical features that characterized the battlefield (i.e. peaks and mountain ranges). The *CI 115 dies* museum is actually the only one that mentions Pàndols and Cavalls, the two mountain ranges where most combats took place and in fact defined the battlefield itself. It can be observed that all museum narratives show a strong bias to minimize the importance of military landmarks beyond towns; this could be explained by the challenge posed by such a

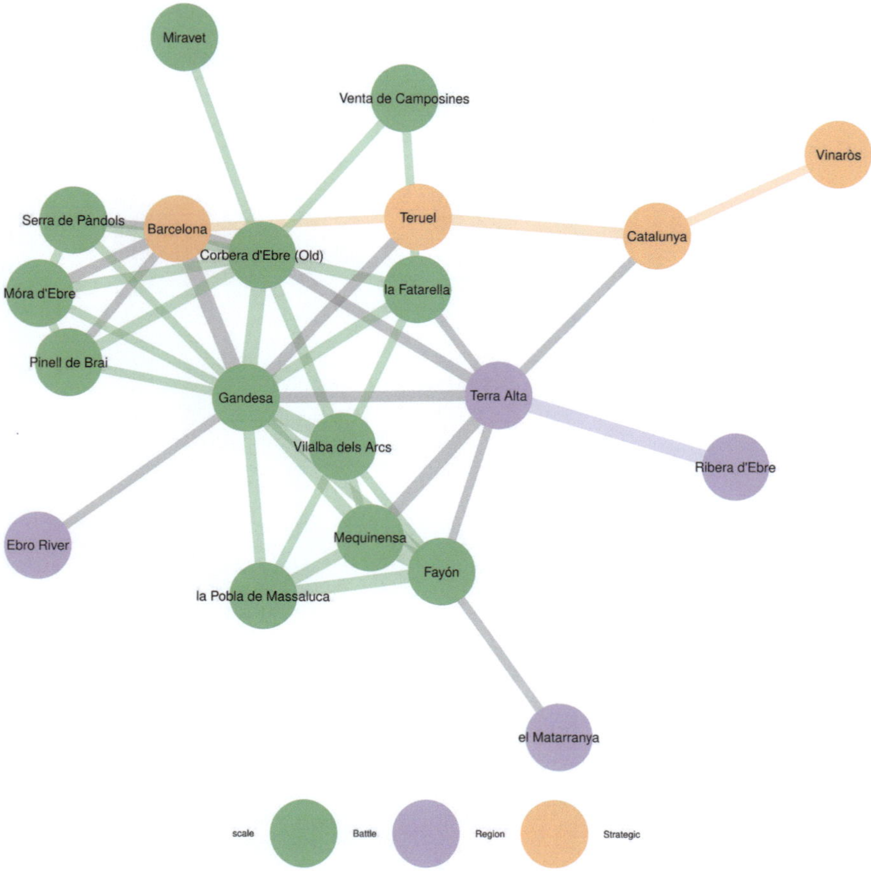

Fig. 4 Network of interactions between locations based on co-occurrences found at least four different text panels. Each node is a location coloured by its scale (i.e. the strategic development of the war, the region where the offensive took place, or battle landmark)

complex battlescape. The topography of Terra Alta is hard to grasp for any visitor because it includes a large number of peaks of similar height that dominate the main natural routes and settlements. The commanders themselves where often confused and troops sometimes assaulted the wrong peaks or took dead-end paths while trying to arrive at the peaks controlled by the enemy (Martínez Reverte 2003, p. 218). This complexity is inherent to any battlescape, but the Ebro is particularly challenging because it was fought on peaks and mountain ranges, and this broken landscape characterized the personal experience of soldiers. Instead, the museum narratives focus on combats in towns and landmarks along the main axis of advance. These axes can be identified in the network of co-occurrences: Venta de Camposines-Corbera d'Ebre-Gandesa and Fatarella-Vilalba dels Arcs.

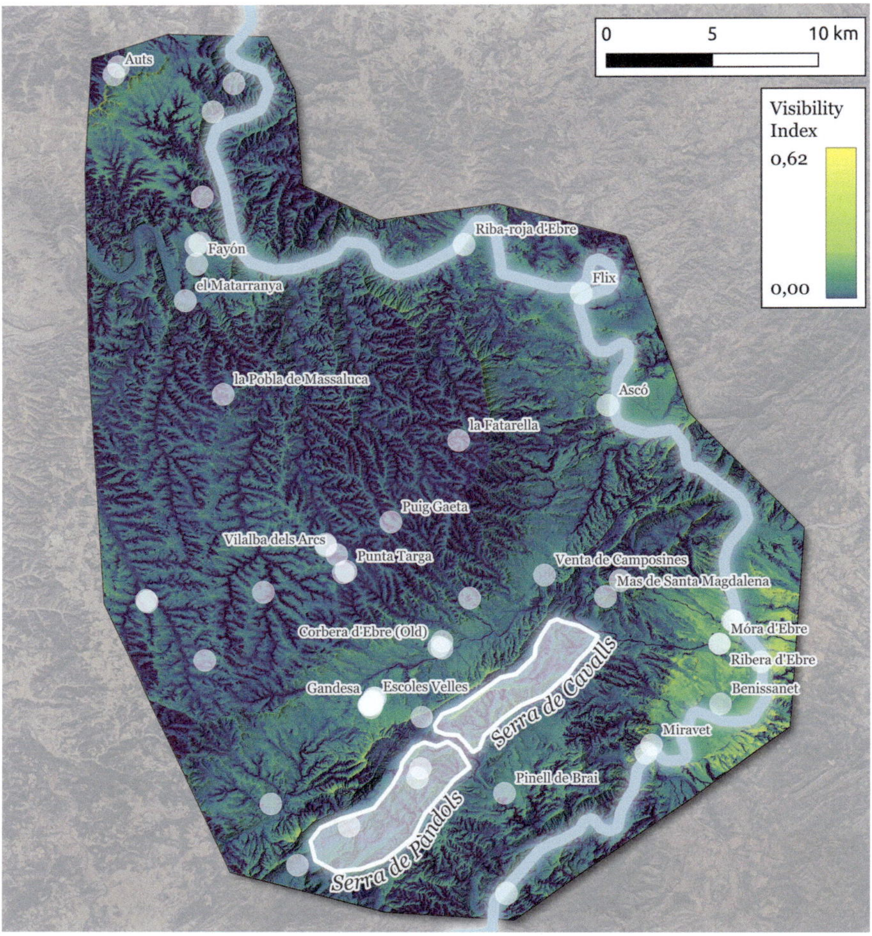

Fig. 5 Visibility index for the entire area of the Ebro battlefield. Labels are shown for places mentioned at least thrice in the dataset

Moving to spatial analysis, the low visibility index values found in the Ebro are one of the main reasons that explained the slow pace of Francoist counteroffensives from a peak towards the next peak. Any of these peaks were excellent defensive positions while also serving as observation points for watching the enemy and guiding indirect fire (e.g. artillery). If the Francoist army took any of these peaks, then the Republicans would defend the next one at the rearguard, and the observation points would only be useful for small viewsheds as the line of sight would be broken by other heights. As a consequence, if an army advanced some hundreds of meters, then enemy positions would visually dominate again the conquered territory, thus forcing another assault to control new observations posts to facilitate further advances. The method devised by Franco to reconquer the territory captured by the Republican

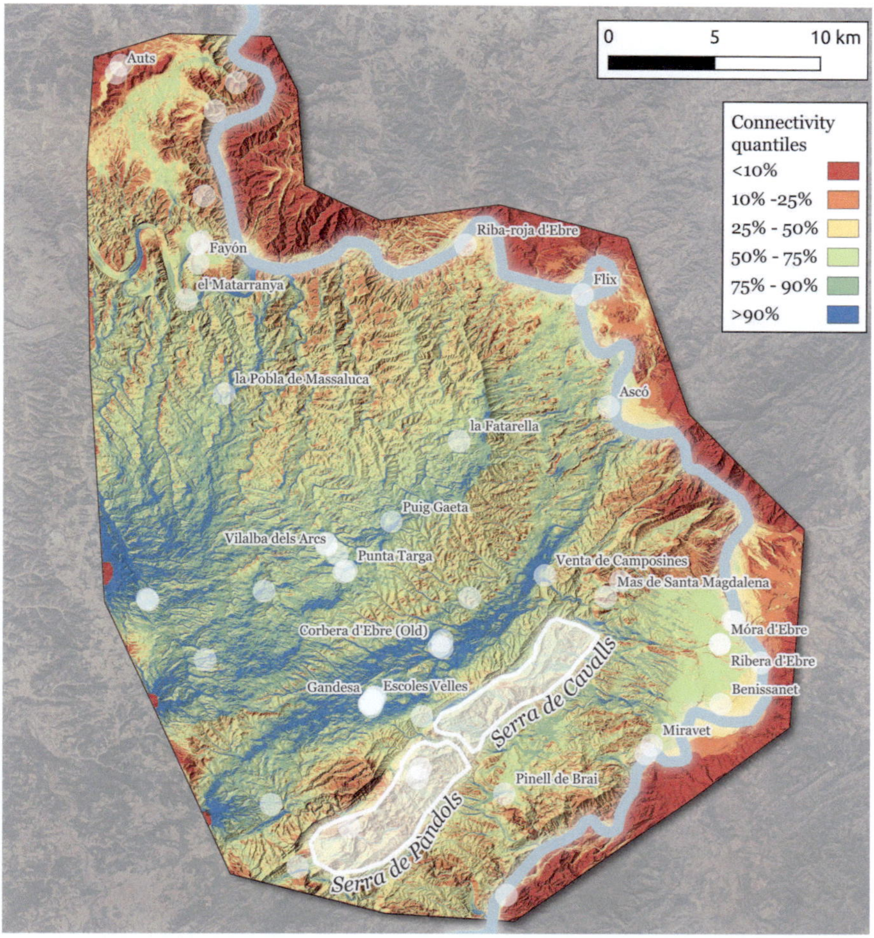

Fig. 6 Connectivity values for the entire area of the Ebro battlefield. The colour gradient is mapping quantiles. Labels are shown for places mentioned at least thrice in the dataset

army can be defined as a loop of massive bombardments followed by costly assaults against the next heavily defended peak. Interestingly enough, this never-ending loop draws strong parallels to military campaigns fought in similar landscapes such as the Allied offensives in Italy during 1943–1944 (Atkinson 2013). Consequently, it can be argued that similar battlescapes saw common tactics and combat dynamics because commanders faced similar geographical challenges and therefore devised resembling strategies regardless of their effectiveness or cost.

At the same time, connectivity analysis highlights the high values of the towns that focused the fight on the valleys (e.g. Gandesa, Corbera d'Ebre); these towns are accompanied by key landmarks that saw bitter struggles for control (e.g. Puig Gaeta, Punta Targa). The scatterplot region with high visibility values and low connectivity

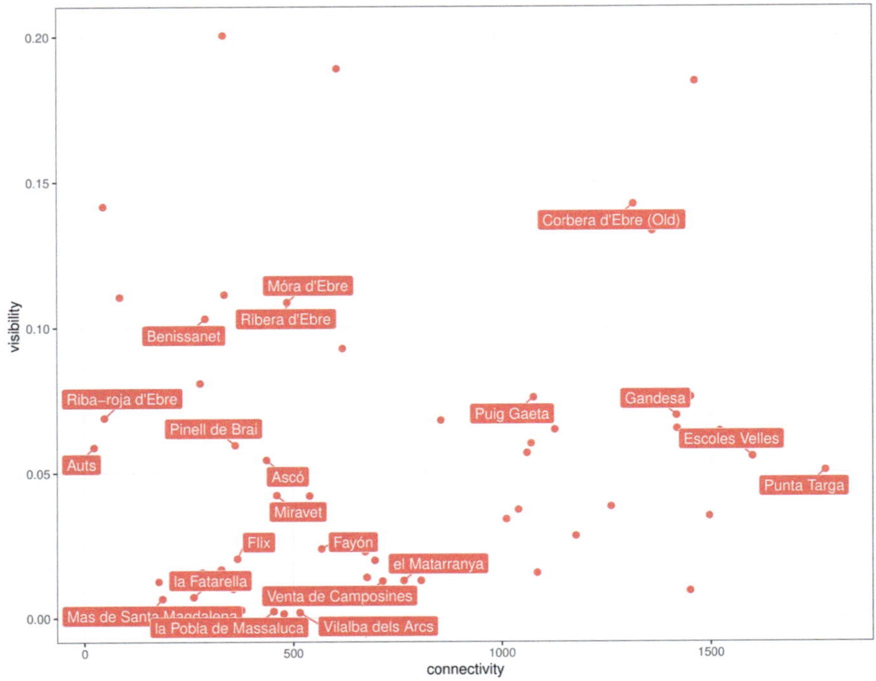

Fig. 7 Scatterplot of locations combining connectivity (X axis) and visibility (Y axis) indexes. Labels are shown for places mentioned at least thrice in the dataset

values includes peaks that were also key combat locations, as several assaults took place to control these points of observations located in key strategic locations for mobility over the battlefield (e.g. Cota 705, Puig de l'Àliga). Another interesting cluster is the set of locations with low visibility and low-medium connectivity such as Vilalba dels Arcs and Faió. These towns were objectives that the Republican army was not able to conquer, and for this reason they became strongholds even if their location was not particularly relevant for battlefield control.

The research approach of this work provides a novel computational framework able to integrate both spatial and text-based analysis of heritage narratives for complex battlescapes. As any other method it has some limitations, being the most important one that museum's exhibit resources that are hard to integrate within the analysis; the content of some interactive tools and videos present in the sample were not included because the work aimed at analysing the main panels of the permanent exhibition. It is worth mentioning that these resources played secondary roles on the examined museums, and its analysis would not change the outcomes in a significant way. A second discarded element were the outdoor panels found in some spots of the battlefield (e.g. Cota 705) which can be analysed in future works to complement the museums following similar methodological strategies (González Cantera and Rubio-Campillo n.d.).

Concluding Remarks

A battle is a historical event where landscape takes a major role given its relevance for combat and regardless of the period or tactics. The Ebro is an excellent example because it was fought for and around landmarks that were not necessarily important beyond the scope of the battle itself. In fact, the assaults to peaks such as the Cim de la Mort or Punta Targa left a terrifying casualty toll while scarring the survivors that participated on them forever. However, these place names do not mean anything to present-day passing visitors to the Terra Alta except for the ones interested on the battle. In the Ebro both armies did not fight for a major city or even a key natural route, but a succession of peaks and villages that happened to be a battlescape precisely because they were difficult to conquer. This work has shown that the importance of these forgotten places is not translated into museum narratives. There is thus a disconnect between the actual heritage and the experience of the battle they purport to explain, precisely because the complexity of the landscape is not effectively communicated to the public.

Is it possible to increase the visibility of geography within the narratives of conflict heritage? A first approach would link the permanent exhibition to the landscape on a stronger way to avoid disconnection between visits to landmarks and what can be seen in the museums. These centres act as the starting point to battlefield visits so their task to explain the landscape is crucial for understanding the battle, but they need appropriate resources for this task. Physical 3D models, large-scale cartography, panorama pictures of points of views, and similar museum components may facilitate the understanding of why these armies were fighting for obscure landmarks, and the huge challenge that this costly fighting generated for the exhausted soldiers and commanders. Second, archaeological research on the battlefield would strengthen the materiality of conflict and its place within a proper landscape context; particularly relevant here seems the role that could be played by the excavation and restoration of trenches, fortified positions, headquarters, and other key locations over the battlefield. It is worth mentioning that archaeology is only mentioned in the panels of the Faió museum, but given the increasing importance of battlefield archaeology it can be argued that it should be better integrated in the rest of the museums and particularly the larger ones (i.e. CI 115 dies and CEBE). Finally, digital heritage technologies could strengthen the link between the museums and the landscape by providing new experiences that establish direct connections between the present-day landscape and the past battlefield. For example, Augmented Reality could show historical pictures at the locations where they were actually taken while Virtual Reality would be useful to reconstruct the landscape of the battle to the visitor, thus generating immersive experiences from the perspective of the soldiers to promote empathy and a better understanding of the spatial dimension of conflict.

The increasing number of battlescapes being promoted as conflict heritage means that there is a need to communicate the importance of these landscapes to present-day society. Beyond military hardware in display cases or documents, landscape is the most important physical heritage remaining from a battle, and it needs to be understood in the context of historical events. This would facilitate a better understanding of the complex interaction between Geography and History, as well as allow visitors to explore the deep imprint that landscape had on the experience of thousands of soldiers and civilians that had to participate in these traumatic events.

Acknowledgements The Digital Elevation Model is based on open 5m. resolution data published by the Instituto Geográfico Nacional and the Institut Cartogràfic and Geogràfic de Catalunya. Connectivity modelling was performed using Circuitscape (Anantharaman et al. 2020) while the Advanced Viewshed Analysis plugin was used for computing the visibility index (Cuckovic 2016). QGIS was used for additional spatial analysis and cartography (QGIS Development Team 2024) and data visualizations were created with R (R Core Team 2024) and ggplot2 (Wickham 2016). All source and data is openly available under open licenses at https://github.com/xrubio/geotextualEbre. The authors warmly thank Cinta Farnós and the rest of the COMEBE, CEBE, and the Faió museum staff for their kind support in data collection and availability.

The project that gave rise to these results received the support of a fellowship from "la Caixa" Foundation (ID 100010434). The fellowship code is LCF/BQ/DR21/11880004. XRC is funded by the Ramón y Cajal programme RYC2018-024050-I (Fondo Social Europeo—Agencia Estatal de Investigación). This research is also part of MAELSTROM PID2023-147060NB-I00 (Ministerio de Ciencia, Innovación y Universidades).

References

Alex B, Grover C, Tobin R, Oberlander J (2019) Geoparsing historical and contemporary literary text set in the City of Edinburgh. Lang Resour Eval 53(4):651–675. https://doi.org/10.1007/s10579-019-09443-x

Alonso Baquer M (1988) La campaña de 1938: Un propósito de nivel político. Ejército. Revista de Las Armas y Servicios 582:55–63

Anantharaman R, Hall K, Shah VB, Edelman A (2020) Circuitscape in Julia: high performance connectivity modelling to support conservation decisions. Proc JuliaCon Conf 1(1):58. https://doi.org/10.21105/jcon.00058

Atkinson R (2013) The day of battle: the War in Sicily and Italy 1943–44. Hachette UK

Burnard L (2013) the evolution of the text encoding initiative: from research project to research infrastructure. J Text Encoding Initiative 5. https://doi.org/10.4000/jtei.811

Busquets C, Ramos J (2019) Perquè el vent ens retorni la memòria! La Batalla de l'Ebre a la llum de l'arqueologia: D'un camp de batalla fossilitzat a un paisatge de Parcs Eòlics. In *Postguerres/Aftermaths of war* (Secció d'Història Contemporània i Món Actual de la Universitat de Barcelona; Ventall Edicions, pp 666–680). https://diposit.ub.edu/dspace/handle/2445/172068

Cairo A (2012) The functional art: an introduction to information graphics and visualization. New Riders

Cao N, Cui W (2016) Introduction to text visualization. Atlantis Press. https://doi.org/10.2991/978-94-6239-186-4

Cardona G, Losada JC (2004) Aunque me tires el puente: Memoria oral de la batalla del Ebro. Aguilar

Cuckovic Z (2016) Advanced viewshed analysis: a quantum GIS plug-in for the analysis of visual landscapes. J Open Source Softw 1(4). https://doi.org/10.21105/joss.00032

Estrada Vidal F (1973) Los que estuvimos en la batalla del Ebro, 2ª ed. E. Jancer; WorldCat

Europe CO (2000) European landscape convention

González Cantera T, Rubio-Campillo X (n.d.) Materialities and memories of the Spanish Civil War: an approach to Bilbao's Iron Belt through digital humanities. Under Review

González Ruibal A (2020) The archaeology of the Spanish Civil War. Routledge, Taylor & Francis Group

González Vázquez D (2016) La patrimonialización de la memoria histórica: Entre el deber social y la estrategia turística. Apuntes sobre el caso catalán. PASOS Revista de Turismo y Patrimonio Cultural 14(5):1267–1280. https://doi.org/10.25145/j.pasos.2016.14.085

Gregory I, Donaldson C, Murrieta-Flores P, Rayson P (2015) Geoparsing, GIS, and textual analysis: current developments in spatial humanities research. Int J Humanit Arts Comput 9(1):1–14. https://doi.org/10.3366/ijhac.2015.0135

Henríquez Caubín J (2009). La Batalla del Ebro: Maniobra de una División. Silente Guadalajara; WorldCat

Howey MCL (2011) Multiple pathways across past landscapes: circuit theory as a complementary geospatial method to least cost path for modeling past movement. J Archaeol Sci 38(10):2523–2535. https://doi.org/10.1016/j.jas.2011.03.024

Jentsch P, Porada S (2021) From text to data. Digitization, text analysis and corpus linguistics. In: Digital methods in the humanities. Challenges, ideas, perspectives, pp 89–128

Kolberg AS (2023) Sound and vision: battlescape analysis: towards a unified method for battlefield archaeology. Primitive Tider 25:53–70. https://doi.org/10.5617/pt.10826

Martínez Bande JM (with Spain) (1988) La batalla del Ebro (Nueva ed). Editorial San Martín

Martínez Reverte J (2003) La batalla del Ebro. Crítica

McLean A, Rubio-Campillo X (2022) Beyond least cost paths: circuit theory, maritime mobility and patterns of urbanism in the Roman Adriatic. J Archaeol Sci 138:105534. https://doi.org/10.1016/j.jas.2021.105534

McRae BH, Dickson BG, Keitt TH, Shah VB (2008) Using circuit theory to model connectivity in ecology, evolution and conservation. Ecology 89(10):2712–2724. https://doi.org/10.1890/07-1861.1

Murrieta-Flores P, Martins B (2019) The geospatial humanities: past, present and future. Int J Geogr Inf Sci 33(12):2424–2429. https://doi.org/10.1080/13658816.2019.1645336

QGIS Development Team (2024) QGIS Geographic Information System [Computer software]. QGIS Association. https://www.qgis.org

R Core Team (2024) R: a language and environment for statistical computing [computer software]. R Foundation for Statistical Computing. http://www.R-project.org/

Rotherham ID (2024) Battlescapes the impact of terrain on military strategy. Pen & Sword Military

Rubio-Campillo X, Hernàndez FX (2015) Combined Arms Warfare in the Spanish Civil War: The Assault on the Republican Defence Line at Fatarella Ridge. J Confl Archaeol 10(1):52–69. https://doi.org/10.1179/1574077315Z.00000000043

Rubio-Campillo X, Ble E, Pujol À, Sala R, Tamba R (2022) A spatial connectivity approach to landscapes of conflict: Julius Caesar and the Assault to Puig Ciutat (NE Iberian Peninsula). J Archaeol Method Theory 29(4):1059–1089. https://doi.org/10.1007/s10816-022-09549-7

Rubio-Campillo X (2023) Trigo, Pólvora Y Datos: El abastecimiento de Barcelona durante el asedio de 1713–1714 bajo la lente de las Humanidades Digitales. Cuadernos de Prehistoria y Arqueología de La Universidad de Granada 33:207–236. https://doi.org/10.30827/cpag.v33i0.27415

Sospedra-Roca R, Hernàndez-Cardona FX, Boj-Cullell I, Hernàndez-Pongiluppi M (2023) Monuments, escultures i museus en els espais històrics de la Batalla de l'Ebre. Les aportacions del Grup «Didáctica del Patrimoni» de la Universitat de Barcelona. Pasado y Memoria 26:451–480. https://doi.org/10.14198/pasado.22790

Thayn JB, Sampeck K, Spaccapaniccia M (2016) Refining Hernando de Soto's route using electric circuit theory and Circuitscape. Prof Geogr 68(4):595–602

Wickham H (2016) ggplot2: elegant graphics for data analysis. Springer, New York. https://ggplot2.tidyverse.org

Geolocating Endangered Heritage: An Educational Strategy to Enhance Geographical Thinking Among Prospective Social Science Teachers

López-Torres Esther, Calle-Carracedo Mercedes, Miguel-Revilla Diego, and De Miguel González Rafael

Abstract Geography education can make use of heritage and socially relevant issues in order to foster geographical and critical thinking and improve social science initial teacher training. This research presents the results of a study that implemented the ArcGIS Survey123 mobile app with a group of 57 Spanish early-childhood education pre-service teachers with the aim of examining their conceptions of heritage and to explore the potential of the use of geolocation tools after an intervention. The study was able to assess prospective teachers' ability to analyze endangered heritage, while also identifying the typology and spatial distribution of heritage elements all across Spain that they selected and photographed during the intervention. Results also helped understand the way participants assessed the importance of heritage elements and the main arguments that they used, as well as the causes of anthropic deterioration that they identified. Prospective teachers' opinions regarding the use of this geolocating tool were also addressed. Finally, a discussion about the possibilities and potential of integrating heritage education and geolocation into social science initial teacher training is presented.

Keywords Geography education · Geolocation · Heritage · Initial teacher training · Geographical thinking

---

L.-T. Esther (✉) · C.-C. Mercedes · M.-R. Diego
Faculty of Education and Social Work, University of Valladolid, Valladolid, Spain
e-mail: esther.lopez.torres@uva.es

C.-C. Mercedes
e-mail: mercedes.calle@uva.es

M.-R. Diego
e-mail: diego.miguel.revilla@uva.es

D. M. G. Rafael
Faculty of Education, University of Zaragoza, Zaragoza, Spain
e-mail: rafaelmg@unizar.es

R. de Miguel González and P. Rivero Gracia (eds.), *Geospatial Technologies for Heritage Education*, Key Challenges in Geography,
https://doi.org/10.1007/978-3-031-96372-8_14

Introduction

Fostering social science prospective teachers' critical thinking skills has been a key concern for educational researchers specializing in history or geography didactics. This has also been seen as an opportunity to integrate concepts linked to historical and geographical thinking (Sharpe et al. 2016), usually in close connection, as well as to promote the best ways to address these ideas in initial teacher training.

Heritage, especially when this concept is understood from a broad perspective, can be combined with the use of geospatial tools for education. These cannot only help to identify and locate heritage elements, but also provide an opportunity to address current or controversial issues adopting an inclusive and integral perspective that might allow to search for potential solutions to contemporary problems. Addressing these issues in teacher training becomes a new challenge that can contribute, at the same time, to the development of teachers' digital competence and to the appreciation of the role of citizenship in the construction of a better world.

Social Science Initial Teacher Training: Connecting Heritage and Citizenship

Since the end of the last century, the scientific community has adopted a broader and more comprehensive concept of heritage. Progressively, the traditional vision that linked it almost exclusively to temporality and preservation of ancient goods, to the aesthetic values of specific historical-artistic periods and to institutional recognition, has been left behind. Instead, a more complex conceptualization, usually in connection with the idea of citizenship, has emerged as much more prominent. Despite being considered a social construction that "can be historically changing, according to new criteria or interests that determine new purposes in new circumstances" (Prats 1997, p. 20), the concept of heritage has been broadened by different approaches. In particular, its symbolic character, as well as its holistic nature, allow the citizenry as a whole (Hall 2023), and people in particular, an essential role to recognize heritage, its promotion and its conservation, usually in close connection with their own particular identities and interests (Fontal and Marín 2018; Fontal 2022).

This has given rise, since the second decade of this millennium, to the development of various research studies on heritage education from an integrative, interactive, participatory and socio-critical perspective (Fontal et al. 2023). In the Spanish context, different research groups have focused on this type of research over the last two decades, making Spain a leading international actor in the scientific production focused on heritage education (Fontal and Ibañez 2017). Specifically, this theme has become the second most important, only behind history education, in social studies didactics, attending to the volume of articles published in Spanish education journals (Gomez-Carrasco et al. 2019). Other international studies also highlight the educational potential of heritage to address socially relevant issues, how to foster historical thinking and historical consciousness, and how to empower citizens to actively participate in building a better world (Rüsen 2004; Martín-Cáceres and Cuenca-López 2015; van Boxtel et al 2016; Molina Puche and Ortuño Molina 2017; González-Monfort 2019).

For social science educators and researchers, the use of heritage has progressively become a valuable teaching resource to go beyond merely conceptual learning, and to promote critical thinking and allow individuals to take part in society. However, this perception of heritage does not necessarily seem to be the prevailing one, neither among textbooks nor among teachers, where there is still a predominant vision of heritage attached to the temporal and monumental, and that becomes more entrenched the higher the educational stage (Martín-Cáceres and Cuenca-López 2015; Monteagudo Fernádez and Oliveros Ortuño 2016). While teachers can be aware of the didactic value of heritage for the development of historical thinking (Molina Puche and Ortuño Molina 2017), a holistic conception of heritage is rarely present in formal education, contrary to what might be desired.

Improving teacher training in heritage education can be crucial to bridge the gap that exists between the theoretical discourse on heritage education and educational practices (Pagès 2000; Fontal et al. 2017; Chaparro and Felices 2019). On the one hand, this requires bringing teachers into contact with specific methodologies for teaching heritage from a holistic, socio-critical perspective (Pinto 2011). On the other hand, however, it might also be necessary to encourage a reflection among educators on the nature of heritage (what it is, where they locate it, its current state, and its relevance to addressing social issues), particularly in initial training, in order to help them to become aware of their own ideas and preconceptions. This could become a necessary first step to detect shortcomings and to design effective educational programs that could help train those teachers who will help build a better society for tomorrow.

GIS Tools for Spatial Understanding of Socially Relevant Problems

Understanding where events happen, why they happen there and not elsewhere, and learning to search for information that explains reality and helps to improve it involves approaching the social sciences, and particularly geography, from a critical perspective that is committed to current issues. For education, this gives academic content meaningful context (Souto 2018; De Miguel et al. 2016) and connects the content to students' lives to foster an understanding of spatial relevance and geographic reasoning (Sharpe et al. 2016), among other key geographic thinking skills. As highlighted by the Lucerne Declaration (Haubrich et al. 2007), these skills can be further strengthened when classrooms incorporate Information and Communication Technologies (ICT).

This declaration, which aimed to highlight the importance of geographical education to address socially relevant issues, and sustainable development in particular, already pointed out the need to train teachers in the use of ICT for teaching geography in a way that helped search for solutions to social challenges. In particular, the Lucerne Declaration also emphasized the didactic use of Geographic Information Systems (GIS) and mobile tools. Specifically, it underlined how the combined use of these technologies could be particularly beneficial for geographic education (Rickles et al. 2017).

The incorporation of geospatial technologies in the classroom has become a major concern for the International Geographical Union (IGU Commission on Geographical Education 2016) as well as for researchers in geography education (De Miguel 2015; De Miguel et al. 2019; De Miguel and De Lázaro 2020). Studies indicate that the use of these technologies requires teachers to have prior training that they often lack, in areas such as cartography (Rodríguez Lestegás et al. 2021) or regarding the use of specific software and hardware (Sebastián López and De Miguel 2017; Sebastián et al. 2023). Even pre-service teachers, despite being mostly younger than in-service educators, struggle to incorporate Geographic Information Technologies into their teaching (Gómez-Trigueros 2020).

Geography education has focused on training teachers to address gaps in their knowledge and skills, an effort that has also resulted in the creation of the European Framework of Digital Competences for Educators (DigCompEdu) (Redecker 2017). In Spain, the Framework of Reference for Digital Competence in Teaching (INTEF 2022) has adapted this competence framework, maintaining a six-area structure to identify teachers' digital competencies. These areas cover essential aspects of teachers' professional practices, establishing three competence blocks: educators' professional and pedagogical skills, and students' digital competencies. From this point of view, introducing GIS to prospective teachers might not only be crucial for developing their geographic thinking skills (Gómez-Trigueros 2020), but also presents an opportunity to promote the acquisition of digital teaching skills through experiential learning (Kolb 1984).

In this regard, the use of GIS has potential to promote spatial understanding and critical thinking, and those applications focused on addressing how social issues can be geolocated are particularly promising (Sebastián López and De Miguel 2017; Klonari et al. 2023). For this reason, training pre-service teachers to identify and locate endangered heritage can foster reflection and promote awareness of their value and the causes of deterioration. This approach can align with four of the key areas of the DigCompEdu framework (Redecker 2017): digital content creation (area 2), teaching and learning processes (area 3), student empowerment (area 5), and student participation in digital problem-solving (area 6). By following the DigCompEdu framework, teacher education can enhance pre-service teachers' pedagogical competencies and their ability to develop digital competence skills with their very own students in the future, especially in social science initial teacher training (Miguel-Revilla et al. 2020).

Purpose of the Study and Research Questions

Attending to the previous ideas, the main aim of this study is to examine pre-service teachers' conceptions of heritage, and to analyze whether geolocation tools, such as the ArcGIS Survey123 app, can make prospective teachers aware of socially relevant issues connected to heritage conservation and their immediate environment. For this reason, the key research questions that guide the current study are as follows:

(1) What heritage elements do prospective teachers spontaneously select in an intervention applied in initial teacher training? What information regarding typology and spatial distribution can be obtained?
(2) How valuable do participants find the heritage elements that were selected? Which criteria do they use?
(3) In which ways did ArcGIS Survey123 help prospective educators to understand, make visible and raise awareness about socially relevant issues and heritage?

Methodology

Due to the nature of this research, focused on analyzing pre-service teachers' ideas and perceptions, the present study adopts a mixed-methods approach, which can be of special use in educational research (Greene 2007). The study included an initial examination of participants' conceptions about heritage through the implementation of a geolocation-based intervention using a mobile app, which allowed researchers to obtain information about its use. After the intervention app, participants were then asked about their ideas about heritage and geolocation tools for educational use. In this regard, this specific study follows a convergent design (Creswell and Plano Clark 2018), in which qualitative and quantitative data were collected in parallel and analyzed in a later stage.

Participants and Design of the Intervention

The sample that was selected for the study included the 57 pre-service teachers enrolled in the bachelor's degree in early-childhood education offered at the Faculty of Education and Social Work of the University of Valladolid (Spain). All participants (between 20 and 21 years of age) were also enrolled in the third-year course *Social Studies Curricular Development*, which focused on geography and history education. Participants provided their informed consent before the intervention in order to collect data for research purposes.

The intervention was designed taking into account the technical framework developed by the research project *Digital competences, learning processes and cultural heritage awareness: quality education for sustainable cities and communities* (PID2020-115288RB-I00), which was financed by the Spanish State Research

Agency and the Spanish Ministry of Science and Innovation. The said intervention was implemented, attending to a series of stages, in order to integrate research and innovative educational practices in early-childhood education and initial teacher training.

During the first phase of the intervention, researchers introduced tools such as Google Earth and Google Maps in some of the sessions of the course with the aim of allowing pre-service teachers to familiarize themselves with geolocation. This allowed them to use these tools to locate and analyze the distribution of elements observed on a map, design itineraries, and recognize them at both ground level (street view) and using satellite view.

In the second phase, participants specifically addressed some of the issues connected to rural areas, as well as potential solutions connected to the preservation of heritage. Pre-service teachers visited Amayuelas de Abajo, an ecological village in the province of Palencia, with the researchers, allowing them to learn first-hand how people lived and the value of covering these issues in education.

For the next phase, pre-service teachers were asked to put into practice some of the notions that they had learnt. Participants were asked to use a mobile app (ArcGIS Survey123) to detect, locate and create awareness about endangered heritage, something that will be described in the following section. After an analysis of heritage elements that were identified and described by pre-service educators, some of the last sessions of the course also allowed researchers to guide participants about the best way to design educational activities for early-childhood education, and applied a final questionnaire in order to examine some of the participants' ideas after the intervention.

Research Instruments and Data Analysis

Data obtained after participants' use of ArcGIS Survey123, a mobile application developed in the framework of the research project, was the main source of information for this study (Sebastián López and De Miguel 2017). Pre-service teachers were asked to make use of this tool to select, identify and take photographs of endangered heritage. Each photograph was then uploaded by participants using their mobile phone, allowing the application to register the geographic coordinates of each location. In addition, participants were also asked to respond to an embedded questionnaire for each of the individual locations and heritage elements that they had selected.

The application was able to compile information about the specific endangered heritage element and its location, but participants were also asked to identify and characterize it (for instance, whether it could be characterized as tangible or intangible heritage. In addition, pre-service teachers also had to indicate the level of importance that they attributed to said element using a 5-point Likert-like scale, which was also supplemented by a qualitative assessment that could also be typed into the application. This allowed for an individual reflection by each participant of some of the reasons why the elements they themselves have identified were considered important. Finally, participants were asked to identify some of the main causes of anthropic deterioration or endangerment of each heritage element.

After the intervention was finished, researchers gathered all the data from the application server and created an SPSS database with all the information. Each location and heritage element was categorized attending to the specific region and location, as well as its typology. Initially, a series of pre-defined elements were used in order to analyze the information, allowing researchers to identify whether the heritage elements were specific constructions, were connected to the natural environment, or were linked to intangible traditions, among many others. In addition, qualitative answers provided by participants were codified using emergent categories, which allowed researchers to interpret whether the arguments presented by participants regarding the value of the elements that were selected were merely descriptive, or were connected to economic, cultural or even historical conceptions about the importance of heritage. Researchers discussed and cross-checked their interpretations during the coding process to promote qualitative reliability (Gibbs 2007).

Finally, as mentioned in the previous section, a final questionnaire was developed in Microsoft Forms in order to analyze pre-service teachers' ideas once the intervention was finished. The questionnaire focused on the experience using the mobile app by also making use of 5-point Likert-like scale items. Data were also collected and exported into an SPSS database for quantitative analysis.

Results

Focusing now on the results that were obtained, the section has been structured in order to follow the three research questions that guide this study. For this reason, the first section will focus on exploring the typology and spatial distribution of the heritage elements that were selected by the participants. The second section will address the way pre-service teachers assessed the value of these elements, attending to both quantitative and qualitative responses. Finally, the third section will analyze participants' perceptions after the intervention regarding the use of the mobile app and endangered heritage.

Typology and Spatial Distribution of Endangered Heritage

After analyzing the responses provided by the 57 participants by making use of the ArcGIS Survey123 mobile application, it was possible to identify a total of 127 heritage elements that were spontaneously selected and photographed by pre-service teachers. After manually classifying each of these elements, the main characteristics and typology of the responses were examined by the researchers in the first place.

In this regard, results indicate that a very notable majority of the participants chose elements that could be characterized as tangible heritage (89% of the total responses), while under 10% of the selections could be described as intangible heritage (Table 1).

The category that appeared more frequently among the selections made by prospective teachers was that of constructions (77.2% of the total responses): participants chose specific monasteries, castles, local cathedrals or churches, old industrial buildings, palaces, or even old houses, among others (Table 2). While this typology, clearly connected with tangible heritage, was very prominent, it is notable that the second one with a higher number of selections was that of traditions. In this case, participants referenced specific religious festivities, such as the *Holy Week* in the city of Valladolid, traditional dances (like the *jota*), or even the local gastronomic tradition.

Other examples selected and photographed by prospective teachers connected to other typologies include elements connected to natural environment (3.9% of the responses) or human-altered natural environment (3.1% of the selections), such as the *Ruta del Cares* in Asturias, *Salto del Usero* in the Region of Murcia, or the *Nerja caves* in Andalusia.

In this regard, the spatial distribution of the selections made by participants provided remarkable results. After individually analyzing each of the submissions,

Table 1 Tangible or intangible heritage selected by participants

	N	%
Tangible heritage	113	89.0
Intangible heritage	12	9.4
N/A or blank	2	1.6
Total	127	100

Note N = number of responses

Table 2 Typology of the heritage elements selected by participants

	N	%
Constructions	98	77.2
Traditions	7	5.5
Natural environment	5	3.9
Transportation	5	3.9
Human-altered natural environment	4	3.1
Open areas	2	1.6
Urban elements	2	1.6
Sculptures	1	0.8
Civil works	1	0.8
Pictorial representations	1	0.8
Transportation and communication infrastructures	1	0.8
Total	127	100

Note N = number of responses

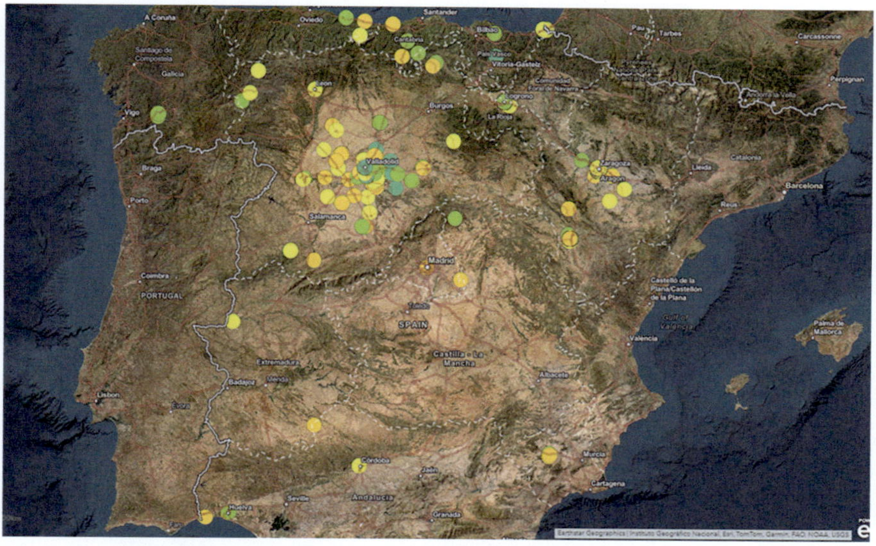

Fig. 1 Spatial distribution of the heritage elements selected and photographed by participants

the researchers classified each heritage element according to the specific geographic coordinates that were sent by the participants and compiled by the mobile application. Prospective teachers were able to use the application in order to send their photographs over their easter holidays, something that allowed many of them to select heritage elements that were not closely connected to the city of Valladolid or even to the region of Castile and León (Fig. 1). This Spanish region grouped 101 of the 127 selections (79.5% of the total number), but the rest of the responses were geolocalized in other regions. In fact, only 64 elements (50.4% of the overall submissions) were found to be located in the province of Valladolid, as many participants opted to photograph elements closer to their birthplace or where their families usually reside.

Instead, a relevant number of heritage elements were located in regions such as Andalusia, Asturias, Galicia or the Basque Country (each of them grouping 3.2% of the responses), while the Region of Murcia, Cantabria, Extremadura and Madrid were also represented. While Aragón also displayed additional results, these were the result of a different intervention in that specific region.

Individual data also allowed the researchers to examine which types of locations related to each of the heritage elements that were photographed. Results indicated that rural areas amounted to 71.7% of the total responses, while elements linked to urban locations were only selected for 26.8% of the submissions (Table 3). As will be discussed later on, the specific conception of heritage found in the participants might have influenced their selection, inviting them to look for tangible or intangible heritage outside cities, and usually focusing on rural areas. On the other hand, this can also be perceived as an inclination to go beyond preconceptions and, sometimes,

Table 3 Rural and urban locations connected to the heritage elements

	N	%
Rural locations	91	71.7
Urban locations	34	26.8
N/A or blank	2	1.6
Total	127	100

Note N = number of responses

even a somewhat limited conception of heritage, in order to value rural areas, their legacy and traditions.

Assessment of Heritage Elements Selected by Participants

For the second research question, the main focus was the assessment of prospective teachers' ideas regarding the value or importance of the heritage elements that they selected. In this regard, the heritage elements selected by the participants were highly valued: over two-thirds of the elements, regardless of their typology, were considered "important" (37.0%) or "very important" (30.7%), while an additional 25.2% were seen as "moderately important." As expected, none of the selected elements were categorized as "not important." However, it's surprising that some of the participants selected heritage elements that they personally viewed as being of "little importance", although these represented only 7.1% of the total (Table 4).

In addition to compiling the assessment for each individual element for data analysis, information was registered on the online ArcGIS platform, allowing to explore the spatial distribution of heritage elements, but also the importance that was attributed by the participant that made the submission (Fig. 2). Each dot presented on the map included a different color representing the level of perceived importance, and included an embedded photograph uploaded by one of the prospective teachers.

More importantly, the participants were asked to explain the reasons why they considered the heritage elements they had spontaneously selected to be valuable. Participants sometimes focused on more than one reason at the same time, but interestingly, for 40.2% of the elements, the participants provided only or in the first-place

Table 4 Assessment of the importance of the selected heritage elements

	N	%
Very important	39	30.7
Important	47	37.0
Moderately important	32	25.2
Slightly important	9	7.1
Not important	0	0
Total	127	100

Note N = number of responses

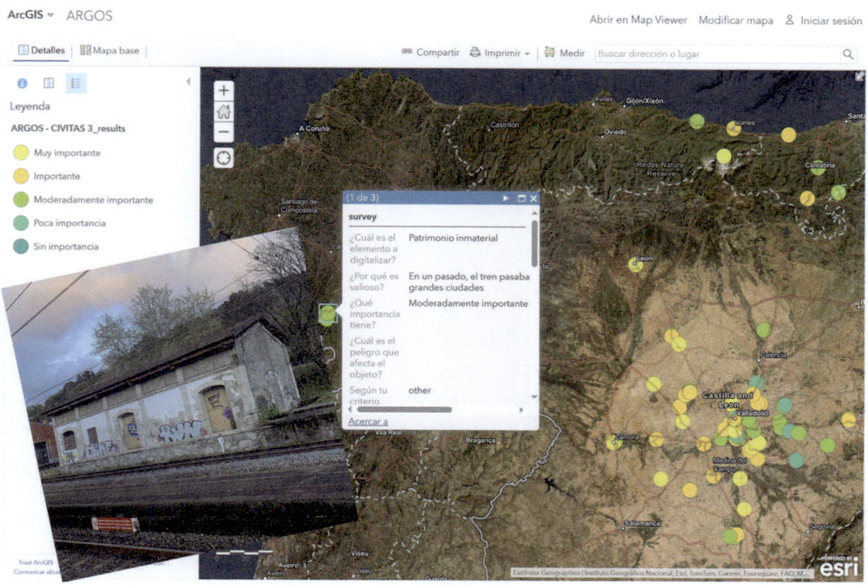

Fig. 2 Screenshot containing an example of a photograph, and responses submitted by a participant. The color of each dot on the map represents the level of importance attributed by each prospective teacher using the Likert-like scale

objective descriptions (including the date of creation, functions, artistic or aesthetic characteristics, authorship, etc.) without expressing any personal interest or judgement. The latter, on the other hand, was found as the first argument in 59% of the remaining responses.

Some arguments are particularly rich and reveal a greater sensitivity to the heritage. For example, in the words of a participant, "the dovecotes found in Tierra de Campos have significant historical and cultural value that should be more appreciated. Mayorga is a historic region known for its agricultural landscape, and the dovecotes are part of its cultural and scenic heritage" (reference: 72). Another example could be found regarding the locomotive depot in the city of Valladolid: "historical value: it witnessed the golden age of the railroad in Spain; architectural value: an example of 19th-century industrial architecture; symbolic value: represents the importance of the railroad in the development of Valladolid" (reference: 84).

Focusing exclusively on the main argument used by participants to assess the importance of the heritage elements that they selected (Table 5), it was possible to detect that those elements valued for their contribution to historical, cultural, and artistic knowledge were the most prominent (30.7%). In some of these responses, it was possible to perceive a conceptualization of heritage as something that can help with the understanding of the past. For example, one student highlighted the value of the railway complex in La Placa (León), noting how "it is valuable because it was once a station that allows us to understand how the steam train, electrification,

Table 5 Categories used to support participants' assessment

	N	%
Descriptive	51	40.2
Historical, artistic, cultural	39	30.7
Symbolic or connected to identities	23	18.1
Practical	10	7.9
Connected to the landscape	4	3.1
Total	127	100

Note N = number of responses

and the transfer of steam worked. On a historical level, it has a fundamental role in post-war energy policy" (reference: 131).

Heritage elements with a strong symbolic and identity-defining character were also abundant in qualitative responses. These elements, usually perceived as representative of a group or a place, encompass both tangible and intangible elements. For example, the Church of the Holy Trinity in Villafuerte de Esgueva is described as "valuable for keeping alive the history, identity and culture of Villafuerte de Esgueva. It is an important religious and cultural center and plays a fundamental role in the municipality, as the majority of the population likes to attend mass" (reference: 104). Similarly, the *jota de las cucharas* is described as "valuable for preserving local identity, strengthening the sense of belonging, and ensuring intergenerational transmission of both the music and anecdotes" (reference: 129).

A small group of heritage elements were considered valuable mainly for their practical usefulness (7.9%). Some participants referenced the specific use that the item has in the present, such as the selection of a seed bank for "seed conservation" (reference: 1), or of fossil trees "because they are educational and cultural" (reference: 130). Other participants, instead, allude to some functions that each element could have in the future, for example, when referring to an uncultivated area, indicating that "trees or any other type of natural element could be planted in this area" (reference: 30). Lastly, a small percentage of participants (3.1%) mentioned aspects connected to the landscape as their first option: locations such as the *Ruta del Cares* in Poncebos (Asturias) (reference: 48), the *Salto del Usero* in Bullas (Murcia) (reference: 88), the *Parque Moret* in Huelva (reference: 123) and a pine forest in Valladolid (reference: 134) were mentioned.

ArcGIS Survey123 to Understand, Make Visible and Raise Awareness About Socially Relevant Issues and Heritage

The third research question focused on the way the use of a geolocating tool such as ArcGIS Survey123 could help prospective teachers to understand and make visible a socially relevant problem. Data obtained from the mobile app, as well as from the questionnaire created using Microsoft Forms were analyzed.

During the intervention, prospective teachers were asked to indicate, using the mobile app, the main reasons why they considered that the heritage element that they had selected and photographed was at risk. In particular, they had to point out whether its deterioration was due to any of the following anthropic causes: lack of conservation, vandalism, inappropriate use, risk of collapse, potential alteration due to urban pressure, etc. Responses provided interesting data that was considered useful to understand the way prospective teachers establish a connection between heritage and human action, from an individual and collective perspective.

Over two-thirds of the heritage elements that were highlighted causes were connected to collective responsibility: for instance, lack of conservation (55.7% of the responses), but also risk of collapse or alteration due to urban pressure (9.7%). While the inappropriate use of heritage elements was selected 10.8% of the time, it might not be clear whether this cause was attributed to an individual or collective responsibility. Other detrimental actions, such as vandalism, were clearly linked to individual interventions and were detected to be important among participants. In fact, vandalism was pointed out by the future teachers as a cause of the deterioration of more than 14% of the heritage elements (Table 6).

Some prospective teachers provided additional reflections on the reasons for the deterioration of heritage elements, beyond those explicitly indicated in ArcGIS Survey123 (the app allowed participants to include an open qualitative answer when they selected *other*). These reflections (10.8% of the selections) pointed to the social changes undergoing in our society, which affect both tangible and intangible heritage. For example, the deterioration of military buildings like the *Conde Ansúrez barracks* in Valladolid was attributed to "the reduction in the size of the Spanish army and the lack of funding for its rehabilitation" (reference: 87). Similarly, traditions like the Holy Week in Valladolid were also mentioned. In the words of one of the participants, "this celebration is not in danger, but we can observe some threats such as socio-cultural changes, the ageing of the brotherhood population, a decrease in participation and some changes in urban dynamics" (reference: 66). The region's gastronomy was also mentioned: "it is in danger because there has been a notable change in the eating habits of the population and in food production, and some culinary knowledge has been lost" (reference: 68).

Table 6 Causes of anthropic deterioration identified by participants

	N	%
Lack of conservation	98	55.7
Vandalism	25	14.2
Risk of collapse or alteration due to urban pressure	17	9.7
Inappropriate use	17	9.7
Other	19	10.8
Total	157	100

Note N = number of responses. This number is 157 instead of 127 due to the fact that some participants selected more than only one option

After the intervention, the analysis of the responses provided by pre-service teachers to the questionnaire implemented in Microsoft Forms allowed to detect how the use of geolocation and of the mobile app allowed them to understand the concept, but also to their awareness regarding heritage. Results indicate that the application was received very positively by participants, and that the type of information provided by the app was also of interest to them (73.6% of the participants assessed this aspect as relevant or very relevant).

Reflecting on the results of the intervention, data show that prospective teachers believe that the mobile app was helpful to them in order to fully understand heritage as a concept (this was described as *very* of *fully* helpful by 52.8% of the participants). Prospective teachers also indicated that using the platform helped them reflect on the diverse typology of heritage, including the distinction between tangible and intangible heritage, as well as between natural and anthropic origins and causes (labeled as *very* of *fully* helpful by 71.7% of them). The application was also described as *very* or *fully* helpful for them to become aware of the diverse spatial location of heritage, including urban and rural areas in 77.3% of the cases, something that was also connected to the high valuation of the app (73.6%) as a tool capable of showing the spatial distribution of different heritage elements.

Finally, the intervention and the use of the ArcGIS Survey123 app was deemed as *very* of *fully* helpful by 56.6% of prospective teachers to comprehend the value of GIS as a way to approach socially relevant issues. In fact, 75.4% of the participants indicated that the use of the application was very or fully helpful in order to make these issues much more visible.

Discussion and Conclusions

After addressing the main research questions presented in this study, it is now possible to establish a discussion of some of the key results that were obtained. In the first place, and in close connection with one of the aims of the study, it should be noted that examining social studies, prospective teachers' ideas about heritage was useful. The use of the ArcGIS Survey123 app in the framework of an intervention allowed researchers to understand the way pre-service educators thought about heritage, and helped establish which ideas and preconceptions were more prominent. In this regard, results indicate that participants' ideas about heritage were not limited to a traditional conception, and that both tangible and intangible heritage were considered, although the fact that tangible heritage was predominantly selected by prospective teachers (in 89% of the submissions), should not be ignored.

This can be, instead, explained by the fact that participants were asked to select, geolocate and photograph a heritage element, which might have motivated them to focus on tangible elements. While this should be considered as a potential limitation, results are in line with previous studies (Martín Cáceres and Cuenca López 2015; Monteagudo Fernádez and Oliveros Ortuño 2016), where the weight of the temporal and monumental was also found to be decisive when thinking about heritage elements, and this was corroborated based on participants' selections (77.2% of them referred to some type of construction). On the other hand, participants were able to recognize different typologies, as well as the spatial diversity of heritage, considering both rural and urban areas, something that might promote a close connection between socially relevant issues, the local environment and social science education.

Secondly, the intervention also allowed to closely analyze in which ways prospective teachers think about heritage and the ways in which they give specific value to heritage elements. Results helped understand the main categories generally connected to value and utility assessment (from the merely descriptive to those focused on historical, artistic, cultural, symbolic or practical ideas). A predominance of simple descriptions was found, usually presenting a simple report, without providing historical, artistic or other types of arguments, although reflections regarding the importance and causes of deterioration helped participants to incorporate a broader vision of heritage. Again, understanding prospective teachers' ideas could be of use to improve teacher education and address the best way to introduce heritage in social science education.

Finally, after analyzing the results obtained during and after the intervention, the use of geolocation tools such as ArcGIS Survey123 mobile app was deemed useful and allowed to foster a more comprehensive vision of heritage, in line with current approaches. Data confirmed the significant potential of integrating GIS and mobile tools for geographic education attending to a contemporary perspective (Rickles et al. 2017). As highlighted by Sebastián López and De Miguel (2017), this application can be a powerful tool to foster geographic and critical thinking, and for acknowledging and acting on current social issues (De Miguel et al. 2016; Souto 2018). As indicated in this study, the ability of making use of spatial analysis in connection with socially relevant issues such as endangered heritage allows for a broader analysis of reality. This can become a reality by introducing new variables that can serve to deepen an understanding of spatial relevance and geographical reasoning (Sharpe et al. 2016).

The use of geolocation in the framework of the intervention has not only allowed to detect preconceptions and ideas, but has also proven to be an interesting tool for trainee teachers to recognize the presence of endangered heritage in rural environments. This finding of the study is particularly significant given the serious problem of depopulation and devaluation facing many rural villages in the region in which the research has been developed. The study reveals that trainee teachers can develop a broader conception of heritage that goes beyond simply its temporal or monumental features, and that they are able to understand it as a social construction (Prats 1997), where spatial location, diverse typologies, as well as identities and symbolism, hold special significance (Fontal and Marín 2018). In this regard, initial teacher training can have an important role in the way prospective educators approach heritage and social science education and can foster a heightened awareness of the potential and many risks facing rural heritage. This, combined with a broad and comprehensive vision, represents an important shift in how pre-service teachers might conceptualize the concept of heritage, and allow for new directions from both a teaching and research perspective.

New research possibilities in the future include a more comprehensive cross-checking between different research instruments, making it possible to explore the different ways in which prospective educators' ideas and preconceptions could be influenced by practices in initial teacher training. Additionally, analyzing if the predisposition of early-childhood education pre-service teachers is similar for other educational stages, such as primary and secondary education, could also be explored in the future. Ultimately, and especially attending to the great potential of some of the approaches that have been described in the study, incorporating geolocation and heritage education into social science teacher training programs can provide teachers with hands-on opportunities to develop geographic and critical thinking skills, something that can also reflect on their teaching practices in the future.

Acknowledgements This research was funded by the project *Digital competences, learning processes and cultural heritage awareness: quality education for sustainable cities and communities* (PID2020-115288RB-I00), financed by the Spanish State Research Agency and the Spanish Ministry of Science and Innovation.

References

Chaparro Á, Felices MM (2019) Perceptions of teachers in initial training about the use of heritage in educational contexts. Revista Interuniversitaria de Formación del Profesorado 94:327–346
Creswell JW, Plano Clark VL (2018) Designing and conducting mixed methods research, 3rd edn. SAGE Publications
De Miguel R (2015) Del pensamiento espacial al conocimiento geográfico a través del aprendizaje activo con tecnologías de la información geográfica. Giramundo 2(4):7–13
De Miguel R, Claudino S, Souto XM (2016) La utopía de la educación geográfica en las declaraciones internacionales de la UGI. In: XIV Coloquio Internacional de Geocrítica. Las utopías y la construcción de la sociedad del futuro. Barcelona
De Miguel R, Donert K, Koutsopoulos K (2019) Geospatial Technologies in Geography Education Springer International Publishing Cham
De Miguel R, De Lázaro ML (2020) WebGIS implementation and effectiveness in secondary education using the digital atlas for schools. J Geogr 119(2):74–85. https://doi.org/10.1080/00221341.2020.1726991
Fontal O, Ibáñez A, Castro PL (2023) La madurez de la educación patrimonial como disciplina: innovar, hibridar, transformar. In: Castro Martín PL, Luna Ú, Martínez M (eds) Tendencias y acciones en educación patrimonial. Dykinson, pp 15–17
Fontal O, Ibáñez A, Martínez M, Rivero P (2017) El patrimonio como contenido en la etapa de Primaria: del currículum a la formación de maestros. Rev Electrónica Interuniv Form del Profr 20(2):79–95. https://doi.org/10.6018/reifop.20.1.286321
Fontal O, Ibañez A (2017) Research on heritage education. Evolution and current state through analysis of high impact indicators. Rev Educ 2017:184–214. https://doi.org/10.4438/1988-592X-RE-2016-375-340
Fontal O, Marín S (2018) Heritage nodes. Analysis of personal heritage links. Arte, Individuo y Sociedad 30:483–500. https://doi.org/10.5209/ARIS.57754
Fontal O (2022) La educación patrimonial centrada en los vínculos. El origami de bienes, valores y personas, Trea. Oviedo
Gibbs G (2007) Analyzing qualitative data. SAGE Publications, London, UK

Gomez-Carrasco CJ, López-Facal R, Rodriguez-Medina J (2019) La investigación en Didáctica de las Ciencias Sociales en revistas españolas de Ciencias de la Educación. Un análisis bibliométrico (2007–2017). Didáctica las Ciencias Exp y Soc, pp 67–88. https://doi.org/10.7203/dces.37.14440

Gómez-Trigueros IM (2020) Digital teaching competence and space competence with TPACK in social sciences. Int J Emerg Technol Learn 15:37–52. https://doi.org/10.3991/ijet.v15i19.14923

González-Monfort N (2019) La educación patrimonial, una cuestión de futuro. Reflexiones sobre el valor del patrimonio para seguir avanzando hacia una ciudadanía crítica. Futur del Pasado 10:123–144. https://doi.org/10.14516/fdp.2019.010.001.004

Greene JC (2007) Mixed methods in social inquiry. Jossey-Bass, San Francisco

Hall S (2023) Whose heritage? Un-settling 'the heritage', re-imagining the post-nation. In: Whose heritage? Challenging race and identity in Stuart Hall's Post-Nation Britain. Taylor and Francis, pp 13–25

Haubrich H, Reinfried S, Schleicher Y (2007) Lucerne declaration on geographical education for sustainable development. Lucerna

IGU Commission on Geographical Education (2016) 2016 International charter on geographical education

INTEF (2022) Marco de Referencia de la Competencia Digital Docente. Madrid

Klonari A, De Lázaro y Torres ML, Kizos A (eds) (2023) Re-visioning geography. Springer International Publishing, Cham

Kolb DA (1984) Experiential learning: experience as the source of learning and development

Martín Cáceres MJ, Cuenca López JM (2015) Educomunicación del patrimonio. Educ Siglo XXI 33:33. https://doi.org/10.6018/j/222491

Miguel-Revilla D, Martínez-Ferreira JM, Sánchez-Agustí M (2020) Assessing the digital competence of educators in social studies: an analysis in initial teacher training using the TPACK-21 model. Australas J Educ Technol 36(2):1–12. https://doi.org/10.14742/ajet.5281

Molina Puche S, Ortuño Molina J (2017) Concepciones del profesorado iberoamericano de Secundaria sobre la contribución del patrimonio local al desarrollo del pensamiento histórico. Estud Pedagógicos XLIII(4):185–202

Monteagudo Fernádez J, Oliveros Ortuño C (2016) The teaching of heritage in the classrooms. An analysis of teaching practices. Unesuniversidad, Escuela y Soc 1:64–79

Pagès J (2000) La formación inicial del profesorado para la enseñanza del patrimonio histórico y de la historia. Treballs d'Arqueologia 6:205–217

Pinto H (2011) Educação Histórica e Patrimonial: conceções de alunos e professores sobre o passado em espaços do presente. Universidade do Minho

Prats L (1997) Antropología y Patrimonio. Ariel, Barcelona

Redecker C (2017) European Framework for the Digital Competence of Educators DigCompEdu. Luxembourg

Rickles P, Ellul C, Haklay M (2017) A suggested framework and guidelines for learning GIS in interdisciplinary research. Geo Geogr Environ 4:2. https://doi.org/10.1002/geo2.46

Rodríguez F, Macía XC, Armas FX (2021) Alfabetización cartográfica: un desafío para la formación del profesorado. Didáctica las Ciencias Experimentales y Sociales 40: 67–80. https://doi.org/10.7203/dces.40.18054

Rüsen J (2004) Historical consciousness: narrative, structure, moral function and ontogenetic development. In: Seixas P (ed) Theorizing historical consciousness. University of Toronto Press, Toronto, pp 63–85

Sebastián M, Kratochvíl O, De Miguel R (2023) Geographic education and spatial citizenship: collaborative mapping for learning the local environment in a global context. In: Re-visioning geography. pp 177–191

Sebastián M, De Miguel R (2017) Educación geográfica 2020. Iberpix y Collector for ArcGis como recursos didácticos para el aprendizaje del espacio. Didáctica Geográfica 18:313–316

Sharpe B, Bahbahani K, Tu Huynh N (2016) Teaching geographical thinking (revised and expanded edition). The Critical Thinking Consortium, Vancouver

Souto XM (2018) School geography: institutional aspirations and classroom experiences. Bol. la Asoc. Geogr. Esp. 2018
van Boxtel C, Grever M, Klein S (eds) (2016) Sensitive pasts. questioning heritage in education. Berghahn Books, New York, NY

The Use of Geolocated Augmented Reality in Teacher Education for the Achievement of the Sustainable Development Goals

Meriam Boulahrouz Lahmidi, Jesús Granados-Sánchez, and Anna Serra Salvi

Abstract This study explores the integration of geolocated augmented reality and the development of the spatial citizenship competence into a Social Sciences Education course of the Degree in Primary Teacher Education of the University of Girona. The aim of this research was to investigate the results of a learning unit that focused on the analysis of the sustainable development goals in the urban landscape. Students were asked to select elements or areas that were relevant for the analysis of sustainability and to formulate actions for improvement. A total of 139 georeferenced entries were created, which were then analysed and categorized using *Atlas.ti 22*. The results of this experience indicate that geospatial technologies promote active learning, critical thinking, and the contextualization of geographic knowledge. Geolocated augmented reality proved to be an effective pedagogical tool for developing spatial awareness and encouraging reflection on sustainability in students' immediate surroundings. However, most of the action proposals were generic in nature, as students formulated simple solutions, and they assigned responsibility to external actors. This leads us to consider the need to develop future teaching guidelines for this learning unit that lead to more concrete and viable action proposals and that focus on the development of students' personal and group agency.

Keywords Geography education · Teacher education · Augmented reality · SDGs · Cultural heritage

M. B. Lahmidi (✉) · J. Granados-Sánchez · A. Serra Salvi
University of Girona, Girona, Spain
e-mail: meriam.boulahrouz@udg.edu

J. Granados-Sánchez
e-mail: jesus.granados@udg.edu

A. Serra Salvi
e-mail: anna.serra@udg.edu

R. de Miguel González and P. Rivero Gracia (eds.), *Geospatial Technologies for Heritage Education*, Key Challenges in Geography,
https://doi.org/10.1007/978-3-031-96372-8_15

Introduction

The 2030 Agenda aims to promote and achieve global sustainable development, and, to this end, it proposes 17 sustainable development goals (SDGs) with specific targets that should be achieved at different spatial scales. Geography education is a discipline that studies most of the issues related to the SDGs and can contribute to their achievement (Granados-Sánchez 2022). In addition, and according to geospatial technologies are essential tools for representation, analysis, projection and modelling of spatial or geographic realities that increase the opportunities to improve the efficiency and effectiveness of the activities we carry out in the field of sustainable development.

In this chapter, we discuss the importance of integrating geospatial technologies, and especially geolocated augmented reality, into initial teacher education with the aim of understanding the vision that future teachers have about the achievements of the SDGs in their environment. Their knowledge should also help them in the design of future possible learning activities on spatial citizenship to be carried out when they start working as teachers.

The chapter is structured in four sections: the first one defines what we understand by geographical competences and the role of geolocated augmented reality in its development and acquisition. The following section describes the context and the educational design of the unit titled: "Analysis of sustainability in urban landscape". Here we present the structure of the learning situation and the activities that students carried out during classroom sessions and fieldwork. The third section briefly explains the methodology used for gathering and the analysis of data. After that, the results section presents the main findings. The chapter ends with a conclusion and discussion of the impact that this unit has had on the students' learning and proposes ways for improvement.

The Use of Geolocated Augmented Reality in Education for the Analysis of SDGs in the Urban Landscape

Geography Education for a Sustainable Global Citizenship

Even though there is no consensus in defining sustainable global citizenship, Granados-Sánchez (2023) has identified the main elements for its characterisation that include: governance, status, the socio-ecological systems, social conscience and engagement. These elements are key for understanding the potential of individual, group and collective agencies for a citizenship as practice (Biesta 2015) or in practice (Van Poeck and Vendenabeele 2013) for the achievement of sustainability.

At the forefront of the transformative journey towards sustainability, there should be educators serving as catalysts for change, shaping the minds of future generations (Ametller et al. 2024). Then, one of the key elements to foster an effective education for sustainability and contribution to the SDGs is the improvement of teachers' capabilities that allows them to act as active change makers in the ecological transition by educating and preparing citizens to cope with future challenges (Serra-Salvi et al. 2019; Serra-Salvi and Medir Huerta 2021).

Teaching and learning about and for the 2030 Global Agenda requires the acquisition of geographical knowledge and competences. Without them, processes such as globalisation, climate change, resource management or migratory movements cannot be fully understood and addressed. On the other hand, geographic information technologies have proved to be tools of paramount use for understanding, analysing and solving problems with a spatial component. These technologies are innovating geography education practices and are contributing to the development of spatial and geographical skills together with the promotion of an active citizenship. Thus, young people can critically interpret and reflect on spatial information and discourses and actively participate in society by making spatial decisions in planning processes (Gryl and Jekel 2012).

De Miguel González (2021) proposes an integrated framework of geographical competences for geography educators articulated in three dimensions: geographical knowledge, geographical skills and geographical practices (see Table 1).

This model has common aspects with the spatial citizen competence (Schulze et al. 2014). The latter focuses on three key elements: use of technology and methodologies; critical analysis; assessment, formulation of proposals and proaction. Therefore, spatial citizen competence enables people to participate in society, to interpret and critically reflect on spatial information, to communicate with the help of maps and other spatial representations, and to express their critical proposals using Geomedia (Schulze et al. 2014) and to negotiate with other stakeholders. These skills are necessary for young people to successfully participate in society and to actively contribute to transforming their society and the entire world.

Table 1 Integrated framework of geographical competences

Geographical knowledge	Geographical thinking	1. Describe: spatial description, geographical patterns and structures in the territory
		2. Explain: physical and human systems. Human–environment interactions social and economic processes. Geographical organisation and settlements
Geographical skills	Spatial thinking	3. Locate: orientation, location, projections, scales
		4. Process: obtaining, processing geographical information, fieldwork, geospatial technologies
		5. Represent: spatial visualisation: text, figures, statistical, cartographical
Geographical practices	Spatial citizenship	6. Interpret: critical thinking, global understanding, spatial imbalances, social justice
		7. Act: intervention, engagement, youth empowerment, social participation, SDGs

Source Adapted from De Miguel González (2021, pages 60 and 64)

Spatial citizenship and the rise of new global communication systems in recent years offer new opportunities for the expansion of citizen education. Digital maps and globes, geographic information systems and augmented reality are reinforcing the interaction of individuals and groups on places and spaces (Gryl et al. 2010). They are also transforming the relationships between physical settings and social circumstances, in which people experience events and phenomena that occur in other parts of the planet and in which they participate. Thanks to these tools, it is easier to train citizens capable of participating and interacting at different spatial scales and to share and promote the SDGs across borders. The didactic implications of that must result in the promotion of examples at different scales, from local to global. Therefore, classroom activities should support active and student-centred learning, integrate multiple perspectives and contexts, and use communication tools that enable participation and negotiation processes through contextualised and cooperative learning environments.

Geolocated Augmented Reality and the Analysis of the Sustainable Development Goals in the Urban Landscape

Augmented reality (AR) is one of the emerging technologies in society, and it is increasingly playing a more relevant role in education (Bacca Acosta et al. 2014). AR is an interactive technology that combines different layers of digital information on physical reality in real time through devices such as tablets or mobile phones, thus creating a new reality enriched with this additional digital information (such as videos, images, text, URLs, pdf files, web pages and so on) (Cabero and Barroso 2016a; Cabero and García 2016).

AR consists of a web platform for the creation and design of content and a mobile app in which the user interacts with the object. The latter allows the capture of the scene, its identification, the increase of information that comes from mixing realities, and finally, it enables the visualisation of the augmented scene (Moralejo et al. 2014). In brief, the augmented reality process is carried out in real time through the mobile app and the capture of a camera, and it establishes a spatial relationship between the virtual information and its real environment. In this way, the user can interact with both real and virtual objects, thus achieving a new communicative space that offers new forms of interactivity.

Geolocated augmented reality offers numerous possibilities and benefits in the educational field, especially for interpreting spaces. These technologies motivate students' interest in their environment, enable interaction while undertaking fieldwork activities and promote meaningful and contextualised learning (Boulahrouz 2021a). The possibility of seeing more information in situ makes this technology one of the most powerful tools of the mobile phone, especially for learning on the ground, which enables new possibilities for the user's experience. AR facilitates action and movement in a physical environment and increases the educational benefit of itineraries on school trips. Nowadays, we can find multiple AR applications, such as Onirix, Layar and Wikitude, to name a few. These types of applications are normally used to make routes around the city or to prepare field trips to places of cultural and/or natural interest. Their purpose is to connect, inquiry and interact with diverse information about those places (whether they are monuments, historical sites and other elements of cultural and natural heritage).

In recent years, these tools have also been used to promote involvement and participation of students in sustainable development (Boulahrouz 2021b). Geolocated AR facilitates the exchange of users' experiences and proposals of action for the improvement of places after spatial citizenship training. It enables young people to reflect on their urban environment and to analyse and evaluate the landscape that surrounds them with sustainability criteria. This technology becomes a potential for connecting the learning acquired in schools with that acquired through informal and non-formal education.

Motiv-ARCHE is a geolocated AR platform that incorporates multiple functionalities (image recognition or 3D, to name a few). It consists of a web application for entering geolocated points and creating content for each of the points, and a mobile app that allows the user to view the previously georeferenced points (Fabregat et al. 2022). This allows the creation of heritage routes and itineraries.

The Motiv-ARCHE application enables users to engage with the following processes:

1. Definition of the heritage element to be added or enhanced.
2. Label the heritage element.
3. Definition of AR triggers: the selection of images for recognition, or to register the geoposition of the heritage element.
4. Associate contents to the heritage element: audios, images, videos, 3D models, web pages, and so on).
5. Label the contents.
6. Define routes or itineraries.
7. Classification of points of interest by its accessibility level and its state of conservation.

Context and Educational Design

This case study focuses on the activity "Analysis of sustainability in the urban landscape". This activity is part of a 6 ECTS course titled "Social Sciences 1" that is part of the second-year study programme of the Degree in Primary Teacher Education of the University of Girona (Spain). The aim of this learning unit is to train future teachers students in the analysis of sustainability based on elements of cultural and natural heritage of the urban landscape using geolocated AR technology. The learning activities that students must undertake consist of choosing points, elements or areas of the urban landscape that are relevant for the analysis of sustainability and to formulate actions that would improve those elements and contribute to the SDGs. To do this, students are asked to employ their spatial skills using geolocated augmented reality. The sequencing of the unit has followed some aspects of the geographical competence framework developed by De Miguel González (2021) and the competency model of spatial citizenship. First of all, it has developed the level of technical and methodological competencies that involve the use of geospatial technologies such as geolocated AR for the development of itineraries and routes. Second, the analysis of the spaces and areas of the urban landscape involved reflection and evaluation processes. Third, participation and communication were encouraged by sharing their proposals with the rest of their classmates.

Motiv-ARCHE was selected as a geolocated AR application for this learning unit. Motiv-ARCHE is a platform for the co-creation of cultural and natural heritage content that was designed by the BCDS[1] research group of the University of Girona. The educational objectives of Motiv-ARCHE are to motivate young people to learn about cultural and natural heritage, making this field of knowledge more attractive and accessible for them, and to facilitate and stimulate the co-creation of heritage content (González Vargas et al. 2021).

The learning unit was designed and structured in five different phases as follows:

1. Introductory phase: The unit started with a short introduction to the aim of the unit and the plan and distribution of sessions of work. The first theoretical input was a lecture about the use of geolocated AR using the Motiv-ARCHE application that was delivered by one of its creators. After that, cooperative working groups of four students were formed and each group chose the municipality of the province of Girona where they wanted to carry out their fieldwork analysis.
2. Fieldwork phase: This working session took place in the field and in non-academic time. During the fieldwork phase, each working group had to identify and select five locations that could be related to the SDGs. In doing so, working groups were free to choose those elements of the landscape they found to be more relevant or appropriate. Participants had to carry out some tasks, such as taking notes and gathering evidence on-site that would describe the location in terms of sustainability, and to take a photograph of the location that would illustrate the situation that was described.
3. Analysis phase: In this phase, students worked in the classroom again. They were entitled to do two activities: the first one consisted of the formulation of two texts: one with the analysis of the place, and another one with an action proposal in order to improve the sustainability of that element. The second activity was the attachment of all evidence to all locations (images and texts).
4. Itinerary phase: After completing the information and evidence of the points of interest, each group defined one itinerary by establishing the order of the points in a route around the map. Students could visualise each element through the virtual platform and the mobile app.

[1] BCDS is the research group on communications and distributed systems of the University of Girona. The members that were involved in the development of Motiv-ARCHE are: Juan Camilo González, Ramón Fabregat, Teodor Jové, Ángela Carrillo

5. Presentation and conclusion phase: In this last phase, each working group of students presented their routes to the other groups. They explained each of the selected elements in terms of their relationship with the SDGs and justified all the actions they proposed in order to improve the achievement of the SDGs in the municipality. The session ended with a discussion about all the experiences and a final reflection and co-assessment of the unit and learning activities.

Methodology

For the purpose of this chapter, we have analysed the open data that was uploaded by students at the Motiv-ARCHE during the development of the unit of study that took place between November 2023 and January 2024. The sample consisted of 139 entries (with text and images, see Fig. 1) created by the working groups formed by 98 s-year students of the degree in Teacher of Primary Education of the University of Girona. Each entry was anonymous, so no personal data was requested. Instead, students used a tag (DCS) for identifying their entries with the learning activity. Students were informed about the possible use of this data for research purposes.

This study used a qualitative research methodology that consisted of documentary analysis using Atlas.ti 24. All selected elements of analysis or points of interest were categorised into nine categories according to the type of space (see Table 2).

The data of each point of interest was analysed following the instrument for the analysis of the spatial competence defined in Table 3.

Table 2 Categories of urban elements and the SDGs

Category	Points of interest	SDGs
Parks and green areas	41	3, 4, 6, 10, 11, 13, 15
Cultural heritage elements	35	4, 8, 9, 11, 17
Squares and streets	21	7, 8, 9, 10, 11
Civic and cultural spaces	11	1, 4, 10, 11
Sports facilities	9	8, 9, 11
Degraded spaces	7	6, 12, 14
Educational centres	6	3, 4, 10
Health care centres	5	3, 11
Social services	4	1, 3, 5,10

Source The authors

Table 3 Instrument for the analysis of the spatial competence

Spatial competence	Geographical competence framework	Elements of analysis
Use of technologies and methodologies	Process: obtaining, processing geographical information, fieldwork, geospatial technologies (4)	1. Use of Motiv-ARCHE app 1.1. Management of the application 1.2. Geolocation of points of interest and creation of routes
Critical analysis	Spatial location, description and explanation (1, 2 & 3)	2.1. Selection of elements
	Critical interpretation (6)	2.2. Problem analysis: What? Where? Why? Link to the SDGs
Assessment, formulation of proposals and proaction	Representation (5)	2.3.Quality of evidence (adequacy, suitability, concret, abstract). Decisions on images and the design of routes
	Action and participation in the SDGs. (7)	3. Quality of suggested action (direct and indirect actions, feasibility, level of specificity and detail) Contribution to the SDGs

Source The authors, using Schulze et al. (2014) Spatial citizenship competence framework and De Miguel González (2021) Geographical competence framework

Findings

The main findings of this study are structured in two parts. The first part presents a general approach to the results following the three elements of the spatial citizenship competence defined in Table 3. The second part focuses on the main aspects related to sustainability and the SDGs that were tackled in the points of interest of each of the nine categories.

General Findings Related to the Spatial Citizenship Competence

The use of Motiv-ARCHE app in this learning unit has proven that students are able to manage a geospatial technology to represent their spatial conceptions about aspects of sustainability of their local environment. Participants consider that the app is very intuitive and enables them to create points of interest and routes with ease. At the same time, the analysis of students' contributions on Motiv-ARCHE demonstrates their technical competence in the use of geolocated AR.

Table 2 shows how the 139 points of interest that were created are classified into 9 spatial categories. All SDGs were included and discussed except goals 2, 16 and 17. The category of parks and green areas is the one with a higher number of entries. The proposed actions were centred on taking care of the environment, the improvement of safety and accessibility to parks, the promotion of quality education and sustainable water management, among others. The second category is made up of cultural heritage elements that include a total of 35 points of interest. This category includes improvements on the renovation of buildings with heritage value, the construction of a more innovative and sustainable infrastructure, the creation of educational programmes and resources in areas of heritage interest and the promotion of sustainable cultural tourism. Squares and streets include 21 points of interest that focus on enabling accessibility, improving mobility in order to avoid traffic congestions, and the prevention and avoidance of risks and dangers in some places. The categories of civic and cultural spaces (11 elements) and cultural centres (6 elements) include actions for the achievement of quality education and the provision of cultural activities according to citizen's interests. Sports facilities, health care centres and social services emphasise the importance of improving buildings, open-air spaces, equipment and increasing the number of professional staff.

The majority of images uploaded to the platform were suitable but generic. Images provide a global vision of each space but they don't zoom in to concrete examples of problems. The pictures from degraded spaces (reporting on uncontrolled landfills) are the only cases where images were specific.

The results show that abstract and/or generic actions predominate, while there are just a few concrete actions. The working groups proposed mainly indirect actions that ascribed the responsibility for implementing the corrective measures to other third parties (primarily local authorities). Although it is true that some actions, such as taking care of the environment, are proposed for citizens, there are just a few actions where students position themselves as protagonists of their implementation. In terms of the object of the actions, they mainly focused on improvements of parks and green areas for educational, health and recreation purposes, accessibility and safety measures, restoration of spaces, and the improvement of certain infrastructures. In the case of heritage sites, they also emphasise the use of these spaces for quality education, promoting innovative infrastructure and improving accessibility, as well as broadening the range of activities for the public to increase their interest in cultural heritage. Another important aspect that they highlight is to restore cultural heritage and attract tourists that would benefit the economic activity in old towns.

Categories and Actions for the SDGs

- **Parks and green areas**: Students selected 41 elements that were related to parks and green areas. This category was the most highlighted. Most of the analysis of these elements had in common the importance of preserving these spaces in good condition, so they can provide the services and functions they have or offer to the people. The points of interest related to parks and green areas are linked to SDGs 3, 4, 6, 11, 13, 15. This association is justified by the fact that green spaces are perceived as fundamental for the preservation of biodiversity (especially by preserving native plants), necessary for promoting human health and personal well-being, as well as a context for quality education activities. The environmental and planning changes that cities have undergone have caused the deterioration of many of these spaces over the years. Actions for the maintenance and conservation of these spaces include raising awareness among citizens and the disposal of waste; to increase the number of litter bins, especially in picnic areas. In relation to SDG 6, they suggest increasing the number of potable water sources and to implement measures to reduce the amount of water consumption:

> This fountain works through a button, which is pressed and water comes out. It consists of a timer that makes the fountain stop flowing after 10 seconds, a mechanism that we believe is very suitable to reduce the water and not leave the fountain flowing all day. However, we think that, in order to save even more water, the water pressure could be reduced.

Accessibility to green spaces is another of the main concerns of the participants. Consequently, they propose the creation of sustainable infrastructures to promote inclusiveness (SDGs 10 and 11):

> It should be noted that the garden may present certain access difficulties for people with reduced mobility or other disabilities. The terrain of this space is irregular, with uneven surfaces and stairs, which makes it difficult for people with mobility difficulties to fully access it. Some areas may be more accessible than others, and some areas may require additional assistance.

- **Elements of cultural heritage**: 35 points of interest about monuments and heritage spaces were georeferenced. These heritage spaces are often related to different SDGs and the importance of their preservation is justified as an element to promote education and culture, but also to boost the economy and promote tourism. From the proposals of action that emerged, it stands out to reopen certain heritage spaces with the aim of revitalising the cultural life of the municipality (SDG 8), and to generate jobs and favour economic growth. Students give educational value to cultural heritage spaces and point out that their preservation is necessary in order to promote quality education (SDG 4). Accordingly, they emphasise that museums should provide more information about what they exhibit and, above all, provide the public with activities related to sustainability that capture their interest:

> The Fishing Museum could promote the importance of reducing maritime pollution and carry out talks so that people become aware of this great problem. In addition, it could do several more practical activities such as cleaning beaches and removing garbage from the sea, among others.

It is considered necessary to improve the accessibility of heritage buildings, as well as to provide them with appropriate furniture.

- **Squares and streets**: The focus of attention of the 21 streets and squares that were analysed was on its general conditions, the preservation of housing, aspects of accessibility, signage, traffic and the promotion of local commerce. As an alternative to avoid traffic and the resulting air pollution, actions are addressed towards increasing public and clean transport facilities, creating wider pedestrian zones, implementing more cycle lanes, promoting the use of electric vehicles and applying taxes to polluting vehicles (SDG 7 and 11). Adapting the urban environment, with the introduction of ramps, lifts and traffic lights with audible signals, would improve accessibility and therefore reduce inequalities (SDG 10).

> To move towards reducing inequalities, it could incorporate sound signals into the traffic light of Jaume I in Girona. It would be an improvement for blind people or people with reduced visibility.

Urban redevelopment accompanied by incentives for the updating and creation of local commerce (SDG 9) would lead to an improvement in the economy and the creation of decent employment (SDG 8).

- **Civic and cultural facilities**: Most of the observations refer to the reasons that lead to underutilisation of spaces. The improvement actions are in line with renewing cultural activities and offering variety, in order to capture the interest of people of different ages and abilities. Creating inclusive spaces would be in line with SDG 10, which seeks to reduce inequalities. Main actions emphasise the priority of taking into account the needs of the local environment and adapting the cultural offer to the demands of the citizenry. These initiatives are considered to be in line with SDG 4 because offering better training and cultural offers are excellent tools to achieve quality education.
- **Sports facilities**: proposals for action include the provision of new sports facilities and the offer of physical activities for the whole population. It is suggested that some abandoned spaces be transformed into sports fields or healthy parks, where people can exercise outdoors as a means of well-being and health promotion. Students pointed out the importance of using energy responsibly and to have water fountains to ensure the hydration of people. They understand that undertaking these environmentally responsible actions is a way to contribute to the construction of a sustainable city (SDG 11) and to help generate employment (SDG 8). At the same time, improving local infrastructure and promoting technological innovation in the sports field helps achieving SDG 9:

> This park could become a park for the implementation of green and innovative technologies, such as efficient lighting, among other aspects. These innovations would not only improve the park, but could also motivate other communities to do the same.

- **Degraded spaces**: the management of waste in certain urban and coastal areas is a higher concern. Main actions are too broad and are focused mainly on prevention, the promotion of responsible production and consumption (ODS 12) and the avoidance of excessive generation of waste. There is a claim for the construction of infrastructures that guarantee a sustainable management of water resources. These initiatives promote water sanitation (SDG 6) and help conserve the seas and their maritime resources (SDG 14):

> It is necessary to improve drinking water infrastructure, promote awareness of responsible water use, implement water-efficient technologies and sustainably manage local water resources.

- **Educational centres**: the condition of buildings and the availability of material and human resources are the main concerns. In small municipalities, there aren't enough teachers and a lack of educational equipment for meeting all the needs of students. Proposals for action include the increase of resources that will result in quality education (SDG 4), reduction of inequalities (SDG 10) and the promotion of children's well-being (SDG 3).
- **Health care centres**: In relation to health facilities, one of the most formulated proposals has to do with the need to encourage the creation of primary health care centres in those populations where there are none or they are very small (SDG 3). The lack of staff in some centres and hospitals is seen as a problem in terms of offering a quality service. The proposals for improvement include systemic actions such as better economic management of hospitals and providing training and incentives for health professionals.
- **Social services**: Although this category is the least referenced, students are committed to implementing social cohesion programmes, promoting education and training in all layers of society and promoting public and inclusive social policies. They consider it essential to improve some social facilities and buildings to dignify people in vulnerable situations. These initiatives can contribute to reducing social inequalities (SDG 10), especially those observed among the elderly and women (SDG 5). The students stressed the lack of involvement by young people in third sector entities. To reverse this situation, they propose to publicise local entities as well as the initiatives they undertake in order to engage younger generations. Finally, we would like to highlight the critical analysis that students carried out regarding some "ghetto neighbourhoods" that suffer from the scarcity of resources that the local authorities allocate for their needs.

Conclusions

The aim of the learning unit "Analysis of the urban landscape" was to develop the spatial citizenship competence among students of the degree in Primary Teacher Education of the University of Girona. Students were involved in the analysis of sustainability in the urban landscape of some localities around Girona (Spain) and presented several proposals for a sustainable action. The results obtained through the use of the geolocated AR application Motiv-ARCHE during the learning unit evidence that the geographical practice carried out allows students to build knowledge based on the observation and analysis of reality (localities were presented as a resource for inquiry and discovery). Participants proposed different actions to promote the SDGs in the local area after activities of critical analysis and reflection.

The development of the learning unit allows us to see that geospatial technologies are very effective for learning in the context of the SDGs. The use of Motiv-ARCHE has proven to be motivating and stimulating for students, and enabled them to link environmental spaces with the promotion of the SDGs in the urban landscape. The use of digital technologies in this context of learning, should be seen as an opportunity to promote active participation, description of responsibility towards the SDGs, and reinforce the importance of the spatial citizenship competence for taking collective action (De Miguel González and Sebastián-López 2022).

It has been verified that the practice has allowed the development of different elements of the spatial citizenship competence: at the level of technical and methodological skills, since the development of the proposal has encouraged students to produce geographic information through the Motiv-ARCHE application; and by reflecting and evaluating the spatial representations of the environment and actively participating with the suggestion of actions (Schulze et al. 2014). This has facilitated the handling of this geospatial tool for the "appropriation of space" lived by manifesting its different points of view and conceptions. In this way, the participants tried to change the meanings of the physical space according to their own points of view and intentions. The descriptions of the points of interest link specific places with different SDGs, thus embracing the interdisciplinary nature of geography. This demonstrates that students have critical knowledge and awareness of the SDGs. The skills of localisation of SDGs in spaces of the close environment together with the analysis and reflection on the different events taking place there, are useful for the training of future teachers. Thus, evidence shows us that they are able to critically analyse aspects of life, identify possible solutions in a creative and innovative way and propose actions for improvement.

Education for sustainable global citizenship requires qualified teachers capable of using geospatial technologies in the creation of participatory and transformative learning activities that promote sustainability. In this study, participants have used geolocated augmented reality and have shown their conceptions about different spaces in their own environment. This allowed them to interact directly in spaces and act as critical and active citizens.

The development of the learning activity shows us that throughout the development process, a series of limitations have been experienced due to the lack of time or the fact that it was implemented for the first time. Future actions for the improvement of this unit of learning include: to move from Motiv-ARCHE (which only produces spatial information on Google Maps) to GIS like ArcGIS or QGIS to scale up the potential of geospatial technologies in heritage and SDG´s education; to provide concrete indications for the development of the images as well as the description and analysis of the spaces; to formulate more detailed proposals for action. This would allow students to really undertake their actions and promote direct and real change in the locality (and not just formulate proposals for action). On the other hand, this training would also help them in the design of future learning activities and projects in schools.

References

Ametller J, Asikainen A, Gual M, Nemejc K (eds) (2024) Teacher training for sustainable development. Developing a shared competence framework. Czech University of Life Sciences, Prague

Bacca Acosta JL, Baldiris Navarro SM, Fabregat Gesa R, Graf S (2014) Augmented reality trends in education: a systematic review of research and applications. J Educ Technol Soc 17(4):133–149

Biesta G (2015) What is education for? On good education, teacher judgement, and educational professionalism. Eur J Educ 50:449–461

Boulahrouz M (2021a) Aprendizaje móvil y ciudadanía espacial en la educación para el desarrollo sostenible: Una propuesta para la enseñanza de las ciencias sociales en la Educación Secundaria Obligatoria. Ediciones Octaedro, Barcelona

Boulahrouz M (2021b) Salidas de campo y educación para el desarrollo sostenible. Una propuesta para la participación juvenil usando el storytelling digital. EDMETIC. Revista de Educación Mediática y TIC 10(2):184–201

Cabero J, y Barroso J (2016a) Posibilidades educativas de la Realidad Aumentada. J New Approaches Educ Res 5:45–52

Cabero J, y García F (2016) Realidad aumentada: tecnología para la formación. Madrid: Síntesis

De Miguel González R, Sebastián-López M (2022) Education on sustainable development goals: geographical perspectives for gender equality in sustainable cities and communities. Sustainability 14(7):4042

De Miguel González R (2021) Geographical and geospatial competences from school education to higher education: the contribution of EUROGEO in the international projects. J Res Didactics Geogr (J-Reading) 2(10):57–65

Fabregat R, Gascons N, Jové T, González-Vargas JC, Coris L (2022) Cocreación de contenidos con Motiv-ARCHE en el proyecto PECT "Costa Brava y Pirineo de Girona: Naturaleza, Cultura e Inteligencia en red". In: Libro de Actas del II Congreso Internacional de Museos y Estrategias Digitales. Servei de Publicacions de la Universitat de València, Valencia, pp 19–28. https://doi.org/10.4995/CIMED22.2022.15649

González Vargas JC, Fabregat R, Carrillo-Ramos A, Jové T (2021) Motiv-ARCHE: an augmented reality application to co-create cultural heritage resources with teenagers. Eur J Post-Classical Archeologies (PCA) 11:387–397

Granados-Sánchez J (2022) Levels of transformation in sustainable curricula: the case of geography education. Sustainability 14:4481. https://doi.org/10.3390/su14084481

Granados-Sánchez J (2023) Sustainable global citizenship: a critical realist approach. Soc Sci 12:171. https://doi.org/10.3390/socsci12030171

Gryl I, Jekel T, Donert K (2010) GI and spatial citizenship. In: Jekel T, Donert K, Koller A (eds) Learning with geoinformation. Wichman Verlag, Berlin

Gryl I, Jekel T (2012) Recentring geoinformation in secondary education: toward a spatial citizenship approach. Cartographica 47(1):2–12

Moralejo L, Sanz C, Pesado P, y Baldassarri S (2014) Avances en el diseño de una herramienta de autor para la creación de actividades educativas basadas en realidad aumentada. Rev Iberoam Tecnol Educ Tecnol 12:8–14

Schulze U, Gryl I, Kanwischer D (2014) Spatial citizenship: creating a curriculum for teacher education. In: GI_Forum 2014—geospatial innovation for society, pp 230–241

Serra-Salvi A, Medir Huerta RM (2021) Enseñar a leer la ciudad en clave de género como punto de partida para una ciudadanía proactiva. In: Granados-Sánchez J, Medir Huerta RM (coords) Enseñar y aprender geografía para un mundo sostenible. Ediciones Octaedro, Barcelona, pp 251–260

Serra-Salvi A, Medir Huerta RM, Servitja Tormo L (2019) Analizar la sostenibilidad de los espacios públicos: estudio de caso para un aprendizaje de ciudadanía global. In: Macía Arce XC, Armas Quintá FX, Rodriguez Lestegas F (coords) La reconfiguración del medio rural en la sociedad de la información. Nuevos desafíos en la educación. Andavira Editora, Lugo, pp 251–260

Van Poeck K, Vandenabeele J (2013) Sustainable citizenship as practice. In: Lifelong learning in Europe, 1–9

Author Index

R. de Miguel González and P. Rivero Gracia (eds.), *Geospatial Technologies for Heritage Education*, Key Challenges in Geography,
https://doi.org/10.1007/978-3-031-96372-8

MIX
Papier aus verantwortungsvollen Quellen
Paper from responsible sources
FSC® C105338

If you have any concerns about our products,
you can contact us on
ProductSafety@springernature.com

In case Publisher is established outside the EU,
the EU authorized representative is:
Springer Nature Customer Service Center GmbH
Europaplatz 3, 69115 Heidelberg, Germany

Printed by Libri Plureos GmbH
in Hamburg, Germany